Luke Robinson

CCEA A2
APPLIED MATHEMATICS

COLOURPOINT
EDUCATIONAL

First Edition
First Impression

Print ISBN: 978 1 78073 347 0
eBook ISBN: 978 1 78073 348 7

Layout and design: April Sky Design
Printed by: GPS Colour Graphics Ltd, Belfast

The Author

Luke Robinson took a mathematics degree, followed by
an MSc and PhD in meteorology. He taught at Northwood
College in London before becoming a freelance mathematics
tutor and writer. He now lives in County Down with his wife
and son.

**COLOURPOINT
EDUCATIONAL**

Colourpoint Educational
An imprint of Colourpoint Creative Ltd
Colourpoint House
Jubilee Business Park
21 Jubilee Road
Newtownards
County Down
Northern Ireland
BT23 4YH

Tel: 028 9182 6339
E-mail: sales@colourpoint.co.uk
Web site: www.colourpointeducational.com

Note: This book has been written to meet the A2
Mathematics specification from CCEA. While the
authors and Colourpoint Creative Limited have taken
all reasonable care in the preparation of this book, it is
the responsibility of each candidate to satisfy themselves
that they have covered all necessary material before
sitting an examination or attempting coursework based
on the CCEA specification. The publishers will therefore
accept no legal responsibility or liability for any errors or
omissions from this book or the consequences thereof.

Contents

Introduction

This book covers the revised specification for Unit A2 2: Applied Mathematics (Mechanics and Statistics) for CCEA, which was available for teaching from September 2018 onwards.

Accuracy

It is important to remember that all answers should be given either exact, or rounded to 3 significant figures. This advice is printed on the front page of all A Level Mathematics papers. Answer marks can be lost for rounding to any other level of accuracy.

Modelling

An important part of Applied Mathematics is **modelling**. Modelling questions may be set in relation to all topics in A2 Applied Mathematics.

What does a modelling question look like?
A modelling question typically involves several of the following features, but not necessarily all of them:

- There may be a requirement to make simplifications. The question will ask what simplifications or assumptions have been made.
- The candidate may be required to discuss the limitations of the model used.
- There may also be a requirement to refine or adapt the model or to consider different models.

The Modelling Cycle
The **Modelling Cycle** is outlined in the diagram. From the wording of the problem, the student should devise a way to model the situation. Simplifications and assumptions may be required.

The model should be applied to obtain a solution and this solution is interpreted and evaluated. At this point, it may become clear that certain assumptions were inappropriate, wrong, or not needed. It may be the case that different assumptions are required. In this way the model can be refined, and this modified version of the model is applied to the problem.

The final report should detail results, conclusions, any assumptions made and any limitations of the model being used. In A2 Level Mathematics, the report will comprise the solution to the problem.

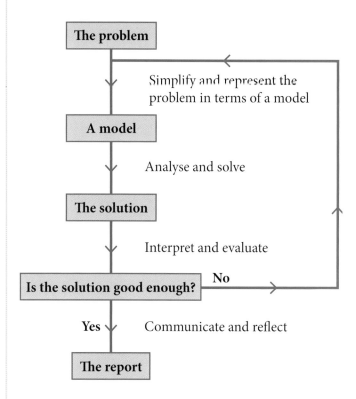

The Modelling Cycle

Chapter 1
Kinematics

1.1 Introduction

Kinematics is the study of the motion of bodies without consideration of the forces that cause them to move. Therefore, in this chapter, there will be no calculations involving forces.

In AS Mathematics you studied kinematics with constant acceleration. In A2 Mathematics, motion with variable acceleration is considered.

Key words
- **Displacement**: The change in position of an object, expressed as a vector.
- **Distance**: The magnitude of the displacement vector.
- **Velocity**: The speed of an object in a given direction. Velocity is a vector quantity.
- **Speed**: The magnitude of the velocity vector.
- **Acceleration**: The rate of change of speed, or the rate of change of velocity.

Before you start
You should:

- Be familiar with the constant acceleration formulae (suvat formulae) and how to use them.
- Be able to apply the differentiation and integration techniques you learnt in AS and A2 Pure Mathematics.
- Understand vectors in two dimensions.

Exercise 1A (Revision)

1. A postman runs away from a dog down a garden path that is 10 metres long. He accelerates at 1 m s^{-2} for 3 seconds. Assuming he starts from rest, determine whether the postman reaches the gate at the end of the path in this time. Show all your working.

Exercise 1A (Revision)

2. $y = 8t^2 + 14 + \dfrac{2}{t}$ where $0 < t \le 4$

   (a) Find $\dfrac{dy}{dt}$ in terms of t.
   (b) Find the value of t that gives a minimum value of y.
   (c) Find this value of y.
   (d) Show that this is a minimum value of y, not a maximum.

3. The diagram shows a triangle PQR.

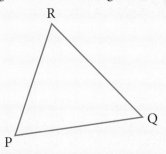

   $\overrightarrow{PQ} = 7\mathbf{i} + \mathbf{j}$, $\overrightarrow{QR} = -5\mathbf{i} + 5\mathbf{j}$ and $\overrightarrow{RP} = -2\mathbf{i} - 6\mathbf{j}$

   (a) Show that the triangle PQR is isosceles.
   (b) Find the area of the triangle.

Notation
- A dot above the name of a variable denotes its first derivative with respect to time. For example, $\dot{\mathbf{r}}$ means $\dfrac{d\mathbf{r}}{dt}$
- Since, in this chapter, the displacement, velocity and acceleration may be functions of time, you may see the notation $s(t)$, $v(t)$ and $a(t)$.

What you will learn
In this chapter you will learn about:

- Motion in a straight line with variable acceleration.
- Motion in 2 dimensions with variable acceleration.

In the real world...

When a rocket is launched, a large part of its mass is the fuel that it carries. As soon as the rocket engines are ignited, fuel is burnt and the overall mass begins to fall.

To work out the rocket's acceleration, Newton's second law $F = ma$ could be rearranged to $a = \dfrac{F}{m}$. This tells us that the acceleration depends on both the mass and the upwards thrust being provided by the engines.

Since the mass is constantly changing as the fuel is burnt, the acceleration is constantly changing as well.

This is an example of **variable acceleration**.

Throughout this chapter you will encounter different systems in which the acceleration is not constant, but is dependent on the time since the object started moving. Typically, to find the distance travelled and the velocity at any time, integration and differentiation are required.

1.2 Motion in a Straight Line With Variable Acceleration

When a body experiences **variable acceleration**, you can model the acceleration as a function of time. You can use calculus to describe the relationship between displacement, velocity and acceleration.

You may have to use any of the functions and techniques from your A-Level pure maths course to analyse motion in a straight line with variable acceleration.

The variables involved in this section are:

s the displacement

v the velocity

a the acceleration

t time

When the acceleration is variable, you cannot use the constant acceleration formulae you learnt in AS Mathematics.

Instead use calculus, as summarised in the following diagram:

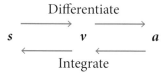

To put this information into equation form:

$$v = \frac{ds}{dt} \qquad \text{and} \qquad a = \frac{dv}{dt} = \frac{d^2 s}{dt^2}$$

$$s = \int v \, dt \qquad \text{and} \qquad v = \int a \, dt$$

For example, to obtain an expression for the velocity, it is possible to differentiate an expression for the displacement with respect to time.

If a question involves integration, remember to include a constant of integration, as shown in the next example. In many cases the question will provide information that will allow you to calculate this constant.

Worked Example

1. A particle moves such that its velocity v m s^{-1} depends upon time t seconds according to
   $$v = t^2 - 3t$$
   The particle is initially at rest at a fixed point O.
   (a) Find an expression for the particle's displacement from O in terms of t.
   (b) Find the particle's displacement from O when $t = 1$.
   (c) Find an expression for the particle's acceleration in terms of t.
   (d) Find the time at which the acceleration is zero.

(a) $s = \displaystyle\int v \, dt$

       $= \displaystyle\int t^2 - 3t \, dt$

       $s = \dfrac{1}{3}t^3 - \dfrac{3}{2}t^2 + c$

The particle is initially at point O. In other words when $t = 0$, $s = 0$ $\therefore c = 0$.

$\therefore s = \dfrac{1}{3}t^3 - \dfrac{3}{2}t^2$

(b) When $t = 1$:

$s = \dfrac{1}{3}(1)^3 - \dfrac{3}{2}(1)^2$

$s = -\dfrac{7}{6}$ m, or -1.17 m (3 s.f.)

> **Note:** Remember displacement can be negative. If a question asks for distance, this is always a positive value. In this case the distance from O would be 1.17 m.

(c) $v = t^2 - 3t$

       $a = \dfrac{dv}{dt}$

$$\therefore a = 2t - 3$$

(d) To find the time at which $a = 0$:
$$2t - 3 = 0$$
$$t = 1.5 \text{ s}$$

...

The following two examples demonstrate how to find the total distance travelled by a particle in a certain time interval. Care must be taken if the particle's velocity is zero at some point within this time interval.

...

Worked Examples

2. A particle P is moving so that its velocity v m s^{-1} after t seconds is given by: $v = 3t^2 - 6t$
   Initially P is at rest and has a displacement of 6 m from a fixed point O.
   (a) Find v when $t = 1$.
   (b) Find an expression for the displacement of P from O at any time t.
   (c) Find the total distance travelled by the particle by the time it returns to its initial position.

(a) When $t = 1$, $v = 3(1)^2 - 6(1) = -3$ m s^{-1}

(b) $$s = \int v \, dt$$
$$s = \int 3t^2 - 6t \, dt$$
$$s = t^3 - 3t^2 + c$$

Initially (when $t = 0$), $s = 6$
$$\therefore 6 = (0)^3 - 3(0)^2 + c$$
$$c = 6$$
Therefore:
$$s = t^3 - 3t^2 + 6$$

(c) The particle's velocity is a quadratic expression:
$$v = 3t^2 - 6t$$
The particle is stationary when $v = 0$ so:
$$3t^2 - 6t = 0$$
$$3t(t - 2) = 0$$
$$t = 0 \text{ or } t = 2 \text{ s}$$

When the particle is stationary at $t = 2$ seconds, it changes direction and begins to move back towards its initial position. Next consider the time at which the particle returns to its initial position. This occurs when $s = 6$:
$$s = t^3 - 3t^2 + 6$$
When $s = 6$:
$$6 = t^3 - 3t^2 + 6$$
$$t^3 - 3t^2 = 0$$
$$t^2(t - 3) = 0$$
$$t = 0 \text{ or } t = 3 \text{ s}$$

Therefore, the particle returns to its initial position when $t = 3$.

When $t = 2$, $s = 2^3 - 3(2)^2 + 6 = 2$ m

So the particle moves 4 metres during the first 2 seconds (from $s = 6$ to $s = 2$ m) and then another 4 metres to return to its starting position in the following 1 second.

In total, the particle travels 8 m in this time.

3. Particle P moves such that its velocity v m s^{-1} at any time t seconds can be calculated using the formula
   $v = -t^3 - 2t^2 + 3t$ for $t \geq 0$
   (a) Find the times at which P is at rest.
   (b) Find the distance travelled by P in the first 2 seconds of its motion.
   (c) Find the time at which P has its maximum velocity.

(a) P is at rest when $v = 0$:
$$-t^3 - 2t^2 + 3t = 0$$
$$t^3 + 2t^2 - 3t = 0$$
$$t(t^2 + 2t - 3) = 0$$
$$t(t - 1)(t + 3) = 0$$
$$t = 0 \text{ or } t = 1 \text{ s}$$
$$(t = -3 \text{ can be rejected})$$

(b) $$s = \int v \, dt$$
$$s = \int -t^3 - 2t^2 + 3t \, dt$$
$$s = -\frac{1}{4}t^4 - \frac{2}{3}t^3 + \frac{3}{2}t^2 + c$$

> **Note:** No initial conditions are given, so the displacement can only be given in terms of c at any time. However, it is possible to work out the distance travelled between two times, as the magnitude of the difference between two displacements.

Calculate the displacement at the start and end times, as well as at the times the particle is stationary.
When $t = 0$, $s = c$
When $t = 1$, $s = \frac{7}{12} + c$
When $t = 2$, $s = -\frac{10}{3} + c$
Distance travelled between $t = 0$ and $t = 1$ is:
$$\left| \left(\frac{7}{12} + c \right) - c \right| = \frac{7}{12}$$
Distance travelled between $t = 1$ and $t = 2$ is:
$$\left| \left(-\frac{10}{3} + c \right) - \left(\frac{7}{12} + c \right) \right| = \frac{47}{12}$$

Total distance travelled in the first 2 seconds is:

$$\frac{7}{12} + \frac{47}{12} = 4.5 \text{ m}$$

> **Note:** An alternative method is to carry out definite integration using limits of 0 and 1 to find the distance travelled between these times, then using limits of 1 and 2.

(c) The maximum velocity occurs when $\frac{dv}{dt} = 0$:

$$v = -t^3 - 2t^2 + 3t$$

$$\frac{dv}{dt} = -3t^2 - 4t + 3$$

When the velocity is at its maximum,

$$-3t^2 - 4t + 3 = 0$$
$$3t^2 + 4t - 3 = 0$$
$$t = 0.535 \text{ s or } t = -1.87 \text{ s (which can be rejected)}$$

$$\frac{d^2v}{dt^2} = -6t - 4$$

When $t = 0.535$, $\frac{d^2v}{dt^2} = -7.21$

$-7.21 < 0$ therefore the particle reaches its maximum velocity when $t = 0.535$ s.

....

You may have to differentiate and integrate a trigonometric function, an exponential function or a logarithmic function.

....

Worked Example

4. A particle's motion in a straight line is modelled such that its acceleration a m s^{-2} at time t seconds is given by $a = \sin 2\pi t$
The particle is initially at rest. Find:
   (a) an expression for the velocity at time t seconds,
   (b) the particle's minimum and maximum velocity,
   (c) the distance travelled in the first 2 seconds.

(a) $v = \int \sin 2\pi t \; dt$

$$v = -\frac{1}{2\pi} \cos 2\pi t + c$$

The particle is initially at rest, so when $t = 0$, $v = 0$

$$\therefore 0 = -\frac{1}{2\pi} \cos(0) + c$$

$$c = \frac{1}{2\pi}$$

So:

$$v = -\frac{1}{2\pi} \cos 2\pi t + \frac{1}{2\pi} \text{ m s}^{-1}$$

(b) The minimum value of $-\frac{1}{2\pi} \cos 2\pi t$ occurs when $\cos 2\pi t = 1$. In this case:

$$v = -\frac{1}{2\pi}(1) + \frac{1}{2\pi}$$
$$= -\frac{1}{2\pi} + \frac{1}{2\pi}$$
$$= 0$$

The minimum velocity is 0 m s^{-1}.

The maximum value of $-\frac{1}{2\pi} \cos 2\pi t$ occurs when $\cos 2\pi t = -1$. In this case:

$$v = -\frac{1}{2\pi}(-1) + \frac{1}{2\pi}$$
$$= \frac{1}{2\pi} + \frac{1}{2\pi}$$
$$= \frac{1}{\pi}$$

The maximum velocity is $\frac{1}{\pi}$ m s^{-1}.

(c) Recall that the distance travelled is the area enclosed between the velocity-time graph and the time axis. Since the particle's minimum velocity is 0 m s^{-1}, the velocity-time graph never goes below the time axis. As such, it is safe to integrate the equation of the curve between 0 and 2 seconds.

$$s = \int_0^2 -\frac{1}{2\pi} \cos 2\pi t + \frac{1}{2\pi} \; dt$$

$$= \frac{1}{2\pi} \int_0^2 -\cos 2\pi t + 1 \; dt$$

$$= \frac{1}{2\pi} \left[-\frac{1}{2\pi} \sin 2\pi t + t \right]_0^2$$

$$= \frac{1}{2\pi} [(0 + 2) - (0 + 0)]$$

$$= \frac{1}{\pi} \text{ or } 0.318 \text{ m (3 s.f.)}$$

> **Note:** If parts of the curve were below the time axis it would be important to find each area above and below the axis separately.

....

Exercise 1B

1. A particle begins at rest at a fixed point O. Its acceleration is given by $a = 5t$, where $t \geq 0$
   Find, in terms of t, expressions for:
   (a) the particle's velocity,
   (b) the particle's displacement from O.

2. At time $t = 0$ a particle P leaves the origin O and moves along the x-axis. At time t seconds the velocity of P is v m s^{-1}, where $v = 8t - t^2$
   (a) Find the maximum value of v.
   (b) Find the time taken for P to return to O.

3. A particle P moves along the x-axis. The acceleration of P at time t seconds, $t \geq 0$, is $(3t + 5)$ m s^{-2} in the positive x-direction. When $t = 0$, the velocity of P is 2 m s^{-1} in the positive x-direction. When $t = T$, the velocity of P is 6 m s^{-1} in the positive x-direction. Find the value of T.

4. A particle moves such that its displacement s metres from a fixed point O at time t seconds is given by $s = 6t^2 - 17t + 5$
   (a) Find the two times t_1 and t_2 at which the particle is at O.
   (b) Find the time at which the particle is at rest.
   (c) Find the distance travelled by the particle between t_1 and t_2.

5. A particle P moves such that its velocity v m s^{-1} at any time t seconds can be calculated using the formula $v = t^3 + 3t^2 - 10t$, for $t \geq 0$
   (a) Find the times at which P is at rest.
   (b) Find the distance travelled by P in the first 3 seconds of its motion.
   (c) Find the time at which P has its minimum velocity.

6. A particle P moves on the x-axis. At time t seconds, its acceleration is $(5 - 2t)$ m s^{-2}, measured in the positive x direction. When $t = 0$, its velocity is 6 m s^{-1} in the positive x direction. Find the time at which P is instantaneously at rest in the subsequent motion.

7. A particle moves in a straight line. At time t seconds after it begins its motion, the acceleration of the particle is $2\sqrt{t}$ m s^{-2} ($t \geq 0$). After 1 second, the particle is moving with velocity $\frac{4}{3}$ m s^{-1}. Find the time taken for the particle to travel 10 m.

Exercise 1B...

8. A yo-yo moves vertically in a straight line so that at time t seconds its acceleration a m s^{-2} is
   $$a = \frac{3}{5}(t - 1) \text{ m s}^{-2} \text{ for } 0 \leq t \leq 5$$
   Initially the yo-yo is at rest.
   (a) Find the **speed** of the yo-yo after 1 second.
   (b) After how many seconds does the yo-yo again come to rest?
   (c) Find how far the yo-yo travels in the first 3 seconds.

9. A particle P moves along a straight line such that its velocity varies according to the formula $v(t) = t^4 - 8t^3 + 17t^2 - 4t$, where $0 \leq t \leq 4.5$ s
   (a) Given that $(t - 4)$ is a factor of $v(t)$, factorise $v(t)$ completely.
   (b) Find the times at which P is stationary.
   (c) Find an expression for the acceleration a in terms of t.
   (d) Given that $a = 0$ when $t = 2$, find the other two times at which the acceleration is 0.
   (e) Find the time at which P has its maximum velocity.

10. A rocket accelerates upwards with an acceleration of $\left(\frac{1}{4}t^2 + \frac{1}{8}t\right)$ m s^{-2}.
    (a) Find the time taken for the rocket's acceleration to reach 10 m s^{-2}.
    (b) Find the distance travelled by the rocket in this time. You may assume that the rocket starts from rest on its launchpad.
    (c) State any modelling assumptions made in the calculations.

1.3 Motion in Two Dimensions With Variable Acceleration

The variables involved in this section are:

s   the displacement vector

v   the velocity vector

a   the acceleration vector

t time (a scalar)

The vector forms of the formulae listed in section 1.2 are as follows:

$$\mathbf{a} = \frac{d\mathbf{v}}{dt} = \frac{d^2\mathbf{s}}{dt^2} \quad \text{and} \quad \mathbf{v} = \int \mathbf{a}\,dt$$

$$\mathbf{v} = \frac{d\mathbf{s}}{dt} \quad \text{and} \quad \mathbf{s} = \int \mathbf{v}\,dt$$

Note: You may see **r** used as an alternative to **s** for the displacement vector.

Differentiation and integration of vectors

Using the above formulae involves differentiation and integration of vector quantities.

To do this, differentiate or integrate each component of the vector separately.

Worked Examples

5. The velocity of a particle at t seconds is given by:
    $$\mathbf{v} = 4t^3\mathbf{i} + 6t^2\mathbf{j}$$
    Given that the particle is initially at a fixed point O, find:
    (a) an expression for the particle's displacement from O at time t,
    (b) an expression for the particle's acceleration at time t.

    (a) $\mathbf{s} = \displaystyle\int \mathbf{v}\,dt = \int 4t^3\mathbf{i} + 6t^2\mathbf{j}\,dt$

    We can integrate each component separately. The vectors **i** and **j** are constants, so:
    $$\mathbf{s} = t^4\mathbf{i} + 2t^3\mathbf{j} + \mathbf{c}$$

    When $t = 0$, the particle is at O, so its displacement from O is $0\mathbf{i} + 0\mathbf{j}$. So:
    $$0\mathbf{i} + 0\mathbf{j} = (0)^4\mathbf{i} + 2(0)^3\mathbf{j} + \mathbf{c}$$
    $$\mathbf{c} = 0\mathbf{i} + 0\mathbf{j}$$
    $$\therefore \mathbf{s} = t^4\mathbf{i} + 2t^3\mathbf{j}\,\text{m}$$

    Note: When integrating a vector, the constant of integration is also a vector quantity.

    (b) Differentiate each component of the velocity vector separately:
    $$\mathbf{v} = 4t^3\mathbf{i} + 6t^2\mathbf{j}$$
    $$\mathbf{a} = \frac{d\mathbf{v}}{dt} = 12t^2\mathbf{i} + 12t\mathbf{j}\,\text{m s}^{-2}$$

    Note: you can also write $\dot{\mathbf{v}}$ for $\dfrac{d\mathbf{v}}{dt}$

Note: Read the question carefully to work out whether you should give a distance or a displacement; a velocity or a speed; a vector acceleration or the magnitude of the acceleration.

6. A particle moves with a velocity $(4t\mathbf{i} + 6t^2\mathbf{j})$ m s^{-1}. Find the magnitude of the acceleration when $t = 2$

    $$\mathbf{v} = 4t\mathbf{i} + 6t^2\mathbf{j}$$
    $$\mathbf{a} = \frac{d\mathbf{v}}{dt} = 4\mathbf{i} + 12t\mathbf{j}$$

    When $t = 2$:
    $$\mathbf{a} = 4\mathbf{i} + 24\mathbf{j}$$

    Since we are asked for the magnitude of the acceleration:
    $$|\mathbf{a}| = \sqrt{4^2 + 24^2} = 4\sqrt{37} = 24.3\,\text{m s}^{-2}\ (3\text{ s.f.})$$

If two particles collide, the **i** components of the two position vectors are equal and the **j** components of the position vectors are equal at the same time. The following example demonstrates how to find the time of collision.

Worked Example

7. The velocity of a particle P at time t seconds (where $t \geq 0$) is $\left((6t^2 - 16)\mathbf{i} + 15\mathbf{j}\right)$ m s^{-1}.
    Initially P is at a point with position vector $(-2\mathbf{i} - 13\mathbf{j})$ m relative to a fixed origin O.
    (a) Find the position vector of P after t seconds.
    (b) A second particle Q moves with constant velocity $(16\mathbf{i} + 12\mathbf{j})$ m s^{-1}. Q is initially at a point whose position vector relative to O is $(-2\mathbf{i} - \mathbf{j})$ m. Show that P and Q collide.

    (a) Let the position vector of P after t seconds be **p**.
    $$\mathbf{p} = \int v\,dt = \int \left((6t^2 - 16)\mathbf{i} + 15\mathbf{j}\right)\,dt$$
    $$\mathbf{p} = (2t^3 - 16t)\mathbf{i} + 15t\mathbf{j} + \mathbf{c}$$

    When $t = 0$, $\mathbf{p} = -2\mathbf{i} - 13\mathbf{j}$
    $$\therefore -2\mathbf{i} - 13\mathbf{j} = 0\mathbf{i} + 0\mathbf{j} + \mathbf{c}$$
    $$\mathbf{c} = -2\mathbf{i} - 13\mathbf{j}$$
    $$\therefore \mathbf{p} = (2t^3 - 16t)\mathbf{i} + 15t\mathbf{j} - 2\mathbf{i} - 13\mathbf{j}$$
    $$\mathbf{p} = \left((2t^3 - 16t - 2)\mathbf{i} + (15t - 13)\mathbf{j}\right)\text{m}$$

    (b) Let the position vector of Q after t seconds be **q**. Since Q moves with constant velocity, we can use the formula:
    $$\mathbf{r} = \mathbf{r_0} + \mathbf{v}t$$
    where $\mathbf{r_0}$ is the initial position vector of the particle and **v** is the constant velocity.

So:

$$\mathbf{q} = (-2\mathbf{i} - \mathbf{j}) + (16\mathbf{i} + 12\mathbf{j})t$$
$$\mathbf{q} = (16t - 2)\mathbf{i} + (12t - 1)\mathbf{j}$$

If P and Q collide, the **i** components of their position vectors must be equal at the same time as their **j** components.

Find the time at which the coefficients of the **i** components are equal for P and Q:

$$2t^3 - 16t - 2 = 16t - 2$$
$$2t^3 - 32t = 0$$
$$2t(t^2 - 16) = 0$$
$$t = 0 \text{ seconds or } t = 4 \text{ seconds.}$$

These are the two times at which the two position vectors have the same **i** components.

For the coefficients of the **j** components:

$$15t - 13 = 12t - 1$$
$$3t = 12$$
$$t = 4$$

At time $t = 4$ seconds, the **i** components and the **j** components of the two position vectors are equal. This is the time at which the two particles collide.

....

The following example requires the times at which a particle is moving parallel to the vector **i**. This occurs when the **j** component of the **velocity vector** is zero. The second part asks for the times at which the particle is due east of the origin. This occurs when the **j** component of its **position vector** is zero.

....

Worked Example

8. A particle moves with an acceleration vector of
    $\mathbf{a} = 12(t^2 - 1)\mathbf{j} \text{ m s}^{-2}, t \geq 0$
    where **i** and **j** are unit vectors east and north of a fixed point O respectively. Initially, P is at the point with position vector $(2\mathbf{i} + 8\mathbf{j})$ m and has velocity $3\mathbf{i}$ m s⁻¹.
    (a) Find the two times at which the particle is moving parallel to the vector **i**.
    (b) Find the two times at which the particle is due east of O.

    _____

    (a) $\mathbf{a} = 12(t^2 - 1)\mathbf{j}$

    $\mathbf{a} = (12t^2 - 12)\mathbf{j}$

    $\mathbf{v} = \int \mathbf{a} \, dt$

    $\mathbf{v} = (4t^3 - 12t)\mathbf{j} + \mathbf{c}$

Initially, P has velocity $3\mathbf{i}$.
$$\therefore 3\mathbf{i} = 0\mathbf{j} + \mathbf{c}$$
$$\mathbf{c} = 3\mathbf{i}$$
$$\mathbf{v} = 3\mathbf{i} + (4t^3 - 12t)\mathbf{j}$$

When moving parallel to **i**, the **j** component of the velocity vector is zero.
$$4t^3 - 12t = 0$$
$$4t(t^2 - 3) = 0$$
$$4t = 0 \text{ or } t^2 = 3$$
$$t = 0 \text{ or } t = 1.73 \text{ s (3 s.f.)}$$
$$(\text{reject } t = -1.73 \text{ s})$$

(b) $\mathbf{r} = \int \mathbf{v} \, dt = \int 3\mathbf{i} + (4t^3 - 12t)\mathbf{j} \, dt$

$\mathbf{r} = 3t\mathbf{i} + (t^4 - 6t^2)\mathbf{j} + \mathbf{c}$

Initially, P has position vector $2\mathbf{i} + 8\mathbf{j}$.
$$\therefore 2\mathbf{i} + 8\mathbf{j} = 0\mathbf{i} + 0\mathbf{j} + \mathbf{c}$$
$$\mathbf{c} = 2\mathbf{i} + 8\mathbf{j}$$
$$\mathbf{r} = 3t\mathbf{i} + (t^4 - 6t^2)\mathbf{j} + 2\mathbf{i} + 8\mathbf{j}$$
$$\mathbf{r} = (3t + 2)\mathbf{i} + (t^4 - 6t^2 + 8)\mathbf{j}$$

When the particle is due east of the origin, the **j** component of the position vector is zero.
$$t^4 - 6t^2 + 8 = 0$$
$$(t^2 - 2)(t^2 - 4) = 0$$
$$t^2 = 2 \text{ or } t^2 = 4$$
$$t = 1.41 \text{ s (3 s.f.) or } t = 2 \text{ s}$$
$$(\text{reject negative solutions})$$

....

Exercise 1C

1. A particle begins at rest at a fixed point O. Its acceleration vector **a** m s⁻² is given by $\mathbf{a} = 12t(\mathbf{i} - \mathbf{j})$, where $t \geq 0$. Find, in terms of t, expressions for:
    (a) the particle's velocity,
    (b) the particle's displacement from O.

2. A particle P starts from rest at a fixed origin O. The acceleration of P at time t seconds (where $t \geq 0$) is $(6t\mathbf{i} + (5 - 3t^2)\mathbf{j})$ m s⁻²
    (a) Find the velocity of P when $t = 4$
    (b) When $t = 2$
        (i) find the position vector of P,
        (ii) find the **distance** of P from O.

3. A particle moves such that its acceleration **a** m s⁻² at time t seconds is given by:
    $\mathbf{a} = 18t^2\mathbf{i} + 24t^2\mathbf{j}$
    Given that the particle begins at rest at a fixed point O, find:

Exercise 1C...

(a) The particle's **speed** after 1 second.

(b) The particle's **distance** from O when $t = \sqrt{2}$ seconds.

4. An ice hockey player accelerates at $t(0.2\mathbf{i} + 0.4\mathbf{j})$ m s^{-2} in the direction $2\mathbf{i} + 3\mathbf{j}$ relative to her goal, where $\mathbf{i}$ and $\mathbf{j}$ are vectors of magnitude 1 m running across and up the rink respectively. She begins at rest, $4\mathbf{j}$ metres from the goal.

   (a) Show that the player's displacement from her goal at time t seconds is given by
   $$\mathbf{r} = \frac{t^3}{30}\mathbf{i} + \left(\frac{t^3}{15} + 4\right)\mathbf{j}\text{ m}$$

   (b) Find the player's distance from her goal after 3 seconds.

   (c) State any modelling assumptions you have made in your calculations.

5. A particle P moves in a horizontal plane. At time t seconds, the position vector of P is $\mathbf{r}$ metres relative to a fixed origin O, and $\mathbf{r}$ is given by $\mathbf{r} = (18t - 4t^3)\mathbf{i} + ct^2\mathbf{j}$ where c is a positive constant. When $t = 1.5$, the speed of P is 15 m s^{-1}.

   (a) Find the value of c.

   (b) Find:
      (i) the acceleration of P, given as a vector in terms of t,
      (ii) the **magnitude** of the acceleration of P when $t = 1.5$

6. A particle is initially at rest at point O. It starts to move from O such that its velocity vector $\mathbf{v}$ m s^{-1} is given by $\mathbf{v} = 3t^2\mathbf{i} - 3t\mathbf{j}$

   (a) Find the acceleration of the particle when $t = 5$.

   (b) Find an expression for the displacement of the particle from O at time t seconds and hence show that the particle does not return to O.

7. At time t seconds (where $t \geq 0$), the velocity of a particle P is $\left((12t - 9)\mathbf{i} + 8\mathbf{j}\right)$ m s^{-1}. Initially P has position vector $(5\mathbf{i} - \mathbf{j})$ m relative to a fixed origin O.

   (a) Find an expression for the position vector of P at time t seconds.

Exercise 1C...

(b) A second particle Q moves with constant velocity $(15\mathbf{i} + \lambda\mathbf{j})$ m s^{-1}. Initially, the position vector of Q is $(35\mathbf{i} + 5\mathbf{j})$ m relative to O. Given that particles P and Q collide, find:
   (i) the time of collision,
   (ii) the value of λ,
   (iii) the position vector of the point of collision.

8. A particle moves with an acceleration vector of $\mathbf{a} = 6(t - 1)\mathbf{j}$ m s^{-2}, for $t \geq 0$ where $\mathbf{i}$ and $\mathbf{j}$ are unit vectors east and north of a fixed point O respectively. Initially, P is at the point with position vector $(-\mathbf{i} - 4\mathbf{j})$ m and has velocity $(2\mathbf{i} + 3\mathbf{j})$ m s^{-1}.

   (a) (i) Show that the particle's velocity is given by $\mathbf{v} = 2\mathbf{i} + A(t^2 - 2t + 1)\mathbf{j}$ m s^{-1}, for $t \geq 0$ where A is a constant to be found.
      (ii) Find the time at which the particle is moving parallel to the vector $\mathbf{i}$.

   (b) (i) Show that the particle's position vector $\mathbf{r}$ m at time t is given by $\mathbf{r} = (Bt + C)\mathbf{i} + (t^3 - 3t^2 + 3t - 4)\mathbf{j}$ where B and C are constants to be found.
      (ii) Show that, if: $(x - 1)^3 = 3$ then: $x^3 - 3x^2 + 3x - 4 = 0$
      (iii) Hence find the time at which the particle is due east of O.

9. A particle moves along a curve such that its displacement vector $\mathbf{r}$ metres varies with time t seconds according to the equation $\mathbf{r} = (2\sin t)\mathbf{i} + (2\cos t)\mathbf{j}$

   (a) Show that the particle remains at a constant distance from the origin O and find this distance.

   (b) Find the initial position vector of the particle.

   (c) Find the velocity of the particle at any time t.

   (d) Show that the **speed** of the particle is constant and find this speed.

Exercise 1C...

10. At time t seconds, a particle has acceleration

$$\left(\frac{(e^t + e^{-t})}{2}\mathbf{i} + \frac{(e^t - e^{-t})}{2}\mathbf{j}\right) \text{m s}^{-2}.$$

The particle starts from rest at a fixed origin O. Show that the particle's displacement from O after the first ln 2 seconds of its motion is

$$\frac{1}{k}(\mathbf{i} - \mathbf{j})\,\text{m}$$

where k is an integer to be determined.

11. A particle moves along the curve shown.

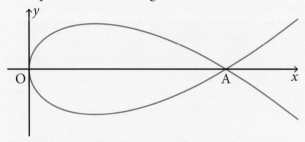

The particle's position vector depends on time according to the formula

$$\mathbf{r} = 3(t-3)^2\mathbf{i} + (3(t-3) - (t-3)^3)\mathbf{j}$$

where $0 \le t \le 6$

(a) Verify that when $t = 3$, the particle passes through the point O.

(b) Find the two times at which the particle passes through point A on the x-axis.

(c) Hence find the position vector of point A.

(d) Show that the **speed** of the particle at any time t is given by

$$\text{Speed} = 3(t^2 - pt + q)$$

where p and q are integers to be determined.

1.4 Summary

A particle's acceleration may be dependent on time. In this situation, the constant acceleration formulae (suvat formulae) cannot be applied to the particle's motion.

Instead, use calculus, as follows:

$$v = \frac{ds}{dt} \qquad \text{and} \qquad a = \frac{dv}{dt} = \frac{d^2s}{dt^2}$$

$$s = \int v\,dt \qquad \text{and} \qquad v = \int a\,dt$$

These formulae can be extended to vector motion as follows:

$$\mathbf{v} = \frac{d\mathbf{s}}{dt} \qquad \text{and} \qquad \mathbf{a} = \frac{d\mathbf{v}}{dt} = \frac{d^2\mathbf{s}}{dt^2}$$

$$\mathbf{s} = \int \mathbf{v}\,dt \qquad \text{and} \qquad \mathbf{v} = \int \mathbf{a}\,dt$$

When integrating remember to include a constant of integration. There will often be enough information in a question to evaluate the constant.

Chapter 2
Projectiles

2.1 Introduction

In mechanics a projectile is an object moving freely under gravity. Its motion may be in one dimension (vertically up and down) or in two dimensions. The case of one dimensional motion was considered in AS Mathematics. In this chapter two dimensional motion of projectiles is considered, with the object moving in the horizontal as well as the vertical direction.

Key words
- **Displacement**: The change in position of an object, expressed as a vector.
- **Distance**: The magnitude of the displacement vector.
- **Velocity**: The speed of an object in a given direction. Velocity is a vector quantity.
- **Speed**: The magnitude of the velocity vector.
- **Acceleration**: The rate of change of speed, or the rate of change of velocity.
- **Range**: The horizontal distance travelled by a projectile.

Before you start
You should:
- Be familiar with the constant acceleration formulae (the 'suvat' formulae) and how to use them.
- Understand vectors in two dimensions.

For an object falling under gravity, if the effects of air resistance are ignored, the object's vertical acceleration is constant. The acceleration does not depend on the object's mass. For example, in a vacuum, an apple and a feather would both accelerate downwards at the same rate. On Earth, the acceleration due to gravity is represented by the letter g. In A-Level Mathematics the approximation $g = 9.8$ m s^{-2} is most often used. Some questions may require a different approximation, e.g. $g = 10$ m s^{-2} or $g = 9.81$ m s^{-2}.

> **Note:** In reality, g is not quite constant. It depends on the object's position on Earth and also its height above sea level.

Exercise 2A (Revision)

1. A particle travels 100 m in a straight line in 4 s. If the particle has a constant acceleration and begins with a velocity of 5 m s^{-1}, find its velocity at the end of its journey.

2. A ball rolls from one side of a table to the other with a constant acceleration of $-0.5\mathbf{i} - 0.25\mathbf{j}$ m s^{-2}. It is given an initial velocity of $4\mathbf{i} + 5\mathbf{j}$ m s^{-1}. If the ball takes 2 seconds to cross the table, find:
   (a) The ball's final displacement relative to its starting position.
   (b) The **distance** travelled.
   (c) The ball's final **speed**.

What you will learn
In this chapter you will learn about:
- Motion under gravity in two dimensions.
- Projectiles.

In the real world...

Tower Bridge in London is a 'bascule bridge', whose two moving sections of road can rise to provide clearance for river traffic.

On 30th December 1952, Albert Gunter was driving the number 78 bus across Tower Bridge, towards Shoreditch. To his utter surprise, the road in front of him seemed to drop away. Gunter quickly realised that Tower Bridge was opening, and his bus was on one of the rising bascules.

Back in the 1950s, a watchman was supposed to ring a warning bell and close the gates before Tower Bridge opened, but on that particular day, he somehow forgot to do this.

Gunter had a choice to make in a split second: would he slam his foot down on the accelerator and hope the bus could jump the gap between the two bascules; or would he slam his foot down on the brake and risk injuring his passengers, or worse still risk skidding off the end of the bascule into the water?

Gunter chose to accelerate and he managed to jump the growing gap between the bascules. He successfully reached the north side of Tower Bridge, which had not yet begun to open, getting all his 20 passengers across safely. For his act of bravery and quick thinking, Albert Gunter was rewarded with a day off work and a reward of £10, about £290 in today's money.

2.2 Modelling Projectile Motion Using The Constant Acceleration Formulae In Two Dimensions

In AS Mathematics you learnt how to use the constant acceleration formulae. These four formulae are shown below.

$$v = u + at$$

$$v^2 = u^2 + 2as$$

$$s = ut + \frac{1}{2}at^2$$

$$s = \frac{1}{2}t(u + v)$$

You can use these formulae to model an object moving vertically under gravity in two dimensions.

Note: In this section, the scalar forms of the constant acceleration formulae are used for the examples and exercise questions. In the following section (2.3) a vector approach is used. However, all of the examples and problems can be solved using either approach. Unless a question specifically requires one of these approaches, it is acceptable to use either.

Horizontal projection
When the initial motion is horizontal, you will often use the constant acceleration formulae in the vertical as well as

$$\text{velocity} = \frac{\text{distance}}{\text{time}}$$

in the horizontal.

In the vertical, it is a good idea to treat the downwards direction as positive, so $a = g = 9.8$ m s^{-2}.

Since the initial motion is horizontal, the initial velocity in the vertical is zero.

In the following example, the vertical motion is considered first to find the time of flight. This is then used in the horizontal to find the distance travelled.

Worked Example

1. Zack's toy gun fires foam pellets. He fires a foam pellet horizontally at Declan with a speed of 10 m s^{-1}. The pellet strikes Declan in the stomach, 40 cm below the height at which it was fired. Find:
   (a) the time taken for the pellet to hit Declan; and
   (b) the distance between the two boys.

(a) Consider the vertical motion to find the time taken. Since the pellet is fired horizontally, the pellet's initial vertical velocity is zero.

$$s = 0.4; u = 0; a = 9.8; t =?$$

$$s = ut + \frac{1}{2}at^2$$

$$0.4 = 0 + \frac{1}{2}(9.8)t^2$$

$$0.4 = 4.9t^2$$

$$t = \sqrt{\frac{0.4}{4.9}} = \frac{2}{7} = 0.286 \text{ s (3 s.f.)}$$

(b) Consider the horizontal motion. The initial horizontal velocity is 10 m s^{-1}. This remains constant throughout the pellet's flight.

$$\text{velocity} = \frac{\text{distance}}{\text{time}}$$

$$\Rightarrow \text{distance} = \text{velocity} \times \text{time}$$

$$\text{distance} = 10 \times \frac{2}{7} = 2.86 \text{ m (3 s.f.)}$$

In the next example, the horizontal motion is considered first to find the time of flight. This is then used in the vertical to find the distance travelled downwards.

Worked Example

2. A dog owner throws a ball horizontally with a speed of 8 m s^{-1} from a height of 1.5 m. The dog runs 4 m horizontally and catches the ball in its mouth. Find the height at which the dog catches the ball.

In the horizontal:

$$\text{velocity} = \frac{\text{distance}}{\text{time}}$$

$$\Rightarrow \text{time} = \frac{\text{distance}}{\text{velocity}}$$

Calculate the time taken for the ball to travel 4 m horizontally:

$$t = \frac{4}{8} = 0.5 \text{ s}$$

In the vertical, $s = ?\,; u = 0\,; a = 9.8\,; t = 0.5$

$$s = ut + \frac{1}{2}at^2 = 0(0.5) + \frac{1}{2}(9.8)(0.5^2) = 1.225\,\text{m}$$

The ball has travelled 1.225 m vertically downwards. Since it started at a height of 1.5 m, the dog catches the ball at a height of 0.275 m or 27.5 cm above the ground.

Exercise 2B

In this exercise use $g = 9.8\ \text{m s}^{-2}$ unless you are instructed otherwise. Give your answers to a suitable level of accuracy.

1. A darts player throws a dart horizontally with a speed of 12 m s⁻¹. The dart hits the dartboard at a point 12 cm below the level at which it was released. Find the horizontal distance travelled by the dart.

2. A tennis ball is served horizontally with an initial speed of 20 m s⁻¹ from a height of 2.5 metres. The net is 1 m high and is situated 12 m horizontally from the server. Determine whether the ball clears the net. Show all your working.

3. A baseball fielder throws the ball horizontally with a speed of 24 m s⁻¹ to the catcher standing 12 m away. If the fielder releases the ball at a height of 2.2 m above ground level find the height of the ball when it reaches the catcher.

4. A policeman in training fires a gun at a target. The bullet initially travels towards the target with a horizontal velocity of 800 m s⁻¹. Given that the target is 20 metres from the policeman, find:
   (a) The vertical displacement of the bullet when it hits the target.
   (b) The **distance** travelled by the bullet.

5. A boy kicks a football horizontally with a speed of 7 m s⁻¹ from the top of a tower block of height 22.5 m.
   (a) Find the football's time of flight.
   (b) Find how far from the base of the tower block the football lands.
   (c) State any modelling assumptions made in your calculations.

6. The top of a vertical tower is 10 m above ground level. When a cannon is fired horizontally from the top of this tower, the cannonball first hits the ground 60 m from the base of the tower. Find the speed with which the cannonball was fired.

7. Two vertical towers of heights 32.4 m and 24.3 m stand on horizontal ground, as shown in the diagram. A ball is thrown horizontally from the top of the higher tower with a speed of 21 m s⁻¹ and just clears the smaller tower.

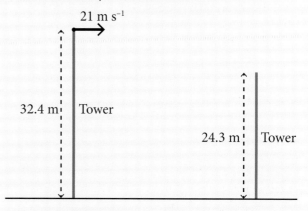

Find the distance
   (a) between the two towers;
   (b) between the smaller tower and the point on the ground where the ball first lands.

8. Point O lies on level ground, at the foot of a tower. Points A and B are on the tower at heights 2.5 m and 14.4 m respectively, as shown in the diagram.

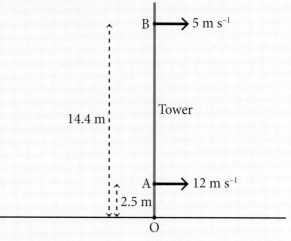

A particle is projected horizontally from B with a speed of 5 m s⁻¹. One second later, a particle

Exercise 2B...

is projected horizontally from A with a speed of 12 m s⁻¹.
(a) Show that the particles reach the ground at the same time.
(b) Show that the particles reach the ground at the same distance from O and find this distance.

9. A and B are two points on level ground. A vertical tower T_1 of height $9h$ metres has its base at A and a vertical tower T_2 of height $4h$ metres has its base at B, as shown in the diagram.

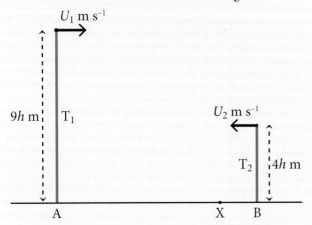

(a) When a stone is thrown horizontally with speed U_1 m s⁻¹ from the top of T_1 towards T_2, it lands at a point X where AX = ⅘AB. When a stone is thrown horizontally with speed U_2 m s⁻¹ from the top of T_2 towards T_1, it also lands at the point X. Show that $3U_1 = 8U_2$.
(b) What modelling assumptions have been made in answering part (a)?

Important results for projectile motion
You will be expected to know the following formulae, which can be derived using the constant acceleration formulae. You may also be asked to prove these results; these proofs are given in Examples 1 and 2 below.

$$\text{Time of flight} = \frac{2u \sin \theta}{g}$$

$$\text{Maximum height} = \frac{u^2 \sin^2 \theta}{2g}$$

$$\text{Range} = \frac{u^2 \sin 2\theta}{g}$$

$$\text{Maximum range} = \frac{u^2}{g}$$

$$\text{Equation of trajectory: } y = x \tan \theta - \frac{gx^2}{2u^2 \cos^2 \theta}$$

Note the following points:

- The formulae for the range and maximum range apply only if the ground is level and the object is projected from ground level. Example 4 below demonstrates the calculation of the range of a projectile that is not projected from ground level.

- The equation of the trajectory is of the form $y = ax + bx^2$ since u, g and θ are constant values. Therefore, the flight path is a quadratic curve and is symmetrical. The time taken for the ascent to maximum height is equal to the time for the descent to ground level. The distances covered are also equal for these two parts of the journey.

- The maximum range occurs when the launch angle θ is 45°.

- The horizontal velocity of the object is $u \cos \theta$. This does not change as the object moves along its trajectory.

- The vertical velocity of the object is initially $u \sin \theta$. This changes according to the constant acceleration formulae.

- Speed at any point on the trajectory is the magnitude of the velocity vector.

The first example below gives a proof of the formulae for time of flight, maximum height, range and maximum range of a projectile.

....

Worked Example

3. A projectile is launched on horizontal ground at an angle of $\theta°$ to the horizontal, as shown in the diagram.

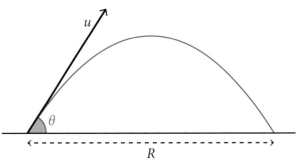

Given that its launch speed is u m s⁻¹, prove that:
(a) The time of flight T of the projectile is given by

$$T = \frac{2u \sin \theta}{g}$$

(b) The maximum height of the projectile is given by

$$H = \frac{u^2 \sin^2 \theta}{2g}$$

(c) The range R of the projectile is given by

$$R = \frac{u^2 \sin 2\theta}{g}$$

(d) The maximum range $R_{\max}$ of the projectile is

given by $R_{\max} = \dfrac{u^2}{g}$

(a) Consider the vertical motion of the projectile for its journey from the ground to its highest point. At the highest point, the vertical component of the velocity is zero. Let the time taken for the journey to this point be T_p. The initial vertical velocity is $u \sin \theta$.

So $u = u \sin \theta$; $v = 0$; $a = -g$; $t = T_p$

For the vertical motion up to the highest point:
$$v = u + at$$
$$0 = u \sin \theta - gT_p$$
$$\Rightarrow T_p = \frac{u \sin \theta}{g}$$

The total time of flight T is $2T_p$, since the upwards and downwards parts of the journey take the same time.
$$\therefore T = \frac{2u \sin \theta}{g} \quad (1)$$

(b) For the maximum height H, again consider the vertical motion to the highest point:

$s = H$; $u = u \sin \theta$; $v = 0$; $a = -g$

$$s = \frac{t}{2}(u + v)$$

$$H = \frac{u \sin \theta}{2g}(u \sin \theta + 0)$$

$$H = \frac{u^2 \sin^2 \theta}{2g}$$

(c) The horizontal component of the velocity of the projectile is $u \cos \theta$. The projectile maintains this horizontal velocity throughout its flight. Since, for constant velocity:

$$\text{velocity} = \frac{\text{distance}}{\text{time}}$$

$$u \cos \theta = \frac{R}{T} \Rightarrow T = \frac{R}{u \cos \theta} \quad (2)$$

From (1) and (2):
$$\frac{2u \sin \theta}{g} = \frac{R}{u \cos \theta}$$

$$\Rightarrow R = \frac{2u^2 \sin \theta \cos \theta}{g}$$

$$R = \frac{u^2 \sin 2\theta}{g}$$

(d) The maximum range occurs when $\sin 2\theta = 1$. So:

$$R_{\max} = \frac{u^2 \times 1}{g} = \frac{u^2}{g}$$

> **Note** that $\sin 2\theta = 1 \Rightarrow \theta = 45°$. Therefore, a projectile travels furthest in the horizontal when launched at an angle of 45° to the horizontal.

The next example proves the formula for the general equation of the trajectory of a projectile.

Worked Example

4. A projectile is launched with initial velocity of u m s^{-1} at an angle of $\theta°$ to the horizontal, as shown in the diagram.

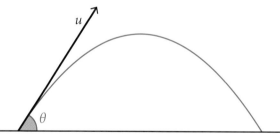

If x and y are the horizontal and vertical distances travelled respectively from the projectile's start point,

show that: $y = x \tan \theta - \dfrac{gx^2}{2u^2 \cos^2 \theta}$

Consider the projectile's journey from the ground to a general point with coordinates (x, y). Let the time taken for this journey be t.

Firstly, consider the vertical motion only. The initial vertical velocity is $u \sin \theta$.

$s = y$; $u = u \sin \theta$; $a = -g$; $t = t$

$$s = ut + \frac{1}{2}at^2$$

$$\Rightarrow y = ut \sin \theta - \frac{1}{2}gt^2 \quad (1)$$

Next consider the horizontal motion. The horizontal component of the velocity is $u \cos \theta$, and this is constant throughout the motion. For the horizontal motion with constant velocity,

$$\text{velocity} = \frac{\text{distance}}{\text{time}}$$

$$\Rightarrow u \cos \theta = \frac{x}{t}$$

$$\Rightarrow t = \frac{x}{u \cos \theta}$$

Substitute into (1):

$$y = u \left(\frac{x}{u \cos \theta} \right) \sin \theta - \frac{1}{2} g \left(\frac{x}{u \cos \theta} \right)^2$$

$$\Rightarrow y = \frac{ux \sin \theta}{u \cos \theta} - \frac{1}{2} g \left(\frac{x^2}{u^2 \cos^2 \theta} \right)$$

$$\Rightarrow y = x \tan \theta - \frac{gx^2}{2u^2 \cos^2 \theta}$$

Note: There are other possible forms for the equation of a projectile's motion, e.g.

$$y = x \tan \theta - \frac{gx^2 \sec^2 \theta}{2u^2}$$

$$y = x \tan \theta - \frac{g}{2} \left(\frac{x \sec \theta}{u} \right)^2$$

$$y = x \tan \theta - \frac{gx^2}{2u^2}(1 + \tan^2 \theta)$$

Example 5 below shows how the formulae for range, time of flight and maximum height can be used.

Worked Example

5. An athlete is taking part in a long jump competition. She propels herself forwards from ground level at an angle of 30° to the horizontal and with a speed of 5 m s^{-1}.
   (a) Find:
       (i) The athlete's horizontal distance travelled.
       (ii) The athlete's maximum height.
       (iii) The time taken for the jump.
   (b) State any modelling assumptions that have been made.

Since the trajectory begins and ends at ground level, the standard formulae can be used.

(a) (i) $\text{Range} = \frac{u^2 \sin 2\theta}{g}$

$$= \frac{5^2 \times \sin(2 \times 30°)}{9.8}$$

$$= 2.21 \, (3 \text{ s.f.})$$

(ii) $\text{Maximum height} = \frac{u^2 \sin^2 \theta}{2g}$

$$= \frac{5^2 \times \sin^2 30°}{2 \times 9.8}$$

$$= 0.319 \text{ m or } 31.9 \text{ cm } (3 \text{ s.f.})$$

(iii) $\text{Time of flight} = \frac{2u \sin \theta}{g}$

$$= \frac{2 \times 5 \times \sin(30°)}{9.8}$$

$$= 0.510 \text{ s } (3 \text{ s.f.})$$

(b) The athlete is modelled as a particle. The effects of air resistance are ignored.

Generalised angle of projection

In Examples 6 and 7 below, the projectile's trajectory does not finish at ground level. In these cases, the standard formulae for horizontal range and time of flight cannot be used. Instead, the constant acceleration formulae can be used in both vertical and horizontal directions.

Worked Examples

6. A football is kicked from a point A at ground level, with a speed of $u = 15.6$ m s^{-1} and at an angle of $\theta°$ to the horizontal, where $\theta = \tan^{-1} \left(\frac{3}{4} \right)$, as shown in the diagram.

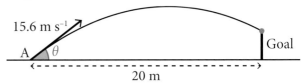

15.6 m s^{-1}

A θ

20 m

Goal

The goal is 20 metres horizontally from point A and the ball is kicked directly towards it, so that its trajectory is at right angles to the goal. The ball hits the crossbar of the goal.
(a) Find the time taken for the ball to hit the crossbar.
(b) Find the height of the goal.
(c) Find the speed of the ball as it hits the crossbar.

Firstly, construct a right-angled triangle to calculate $\cos \theta$ and $\sin \theta$.

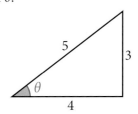

5

3

θ

4

Since $\tan\theta = \dfrac{3}{4}$, the opposite and adjacent sides of the triangle are 3 and 4 respectively. The hypotenuse is calculated as 5 using Pythagoras' Theorem.

From the triangle, $\cos\theta = \dfrac{4}{5}$ and $\sin\theta = \dfrac{3}{5}$.

The horizontal component of the initial velocity is:

$$u\cos\theta = 15.6 \times \dfrac{4}{5} = 12.48 \text{ m s}^{-1}$$

The vertical component of the initial velocity is:

$$u\sin\theta = 15.6 \times \dfrac{3}{5} = 9.36 \text{ m s}^{-1}$$

(a) In the horizontal:

$$\text{velocity} = \dfrac{\text{distance}}{\text{time}}$$

$$\Rightarrow 12.48 = \dfrac{20}{t}$$

$$\Rightarrow t = \dfrac{20}{12.48}$$

$$= 1.60256\ldots = 1.60 \text{ s (3 s.f.)}$$

(b) In the vertical:

$$s = ?;\ u = 9.36;\ a = -9.8;\ t = 1.60256$$

$$s = ut + \dfrac{1}{2}at^2$$

$$s = 9.36 \times 1.60256 + \dfrac{1}{2}(-9.8)(1.60256)^2$$

$$= 2.42 \text{ m (3 s.f.)}$$

At this time, the ball has a vertical height of 2.42 m, so this is the height of the crossbar, i.e. the height of the goal.

(c) To find the speed of the ball as it hits the crossbar, calculate the horizontal and vertical components of the velocity at this time.

The horizontal component of velocity is 12.48 m s^{-1}, since this remains constant along the ball's trajectory.

In the vertical:

$$u = 9.36;\ v = ?;\ a = -9.8;\ t = 1.60256$$

$$v = u + at$$

$$= 9.36 - 9.8 \times 1.60256$$

$$= -6.345088 \text{ m s}^{-1}$$

Draw a velocity vector diagram:

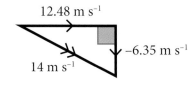

12.48 m s^{-1}

−6.35 m s^{-1}

14 m s^{-1}

$$\text{Speed} = \sqrt{12.48^2 + (-6.345088)^2}$$

$$= 14 \text{ m s}^{-1}$$

7. A ball is thrown with speed 7 m s^{-1} from a window W, which is 3.5 m vertically above horizontal ground. The ball is projected at an angle $\theta°$ to the horizontal. The point P is the highest point on the path of the ball and is 2 m above the level of W. The ball moves freely and hits the ground at the point G, as shown in the diagram.

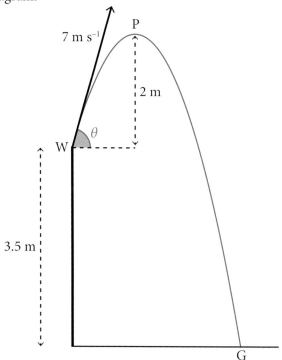

7 m s^{-1}

P

2 m

θ

W

3.5 m

G

Find:

(a) The value of θ.

(b) The horizontal distance travelled by the ball.

(c) The **speed** of the ball as it hits the ground at G.

Use $g = 9.8$ m s^{-2} throughout and give all answers to 1 decimal place.

(a) Use the formula for the maximum height of the ball.

$$\text{Maximum height} = \dfrac{u^2\sin^2\theta}{2g}$$

$$2 = \dfrac{7^2 \times \sin^2\theta}{2g}$$

$$\Rightarrow \sin\theta = \sqrt{\dfrac{4g}{49}} = 0.894\ldots$$

$$\theta = \sin^{-1}(0.894\ldots) = 63.4°$$

(b) Equation of trajectory: $y = x\tan\theta - \dfrac{gx^2}{2u^2\cos^2\theta}$

$$\sin\theta = \sqrt{\dfrac{4g}{49}} = \dfrac{2\sqrt{5}}{5}$$

$$\Rightarrow \cos\theta = \dfrac{\sqrt{5}}{5} \text{ and } \tan\theta = 2$$

(Finding $\cos\theta$ and $\tan\theta$ from $\sin\theta$ is left as an exercise for the reader.)

$$\therefore y = 2x - \frac{9.8x^2}{2 \times 7^2 \times \left(\frac{\sqrt{5}}{5}\right)^2}$$

$$y = 2x - \frac{1}{2}x^2$$

This is the equation of the trajectory in its simplest form. To find the horizontal distance travelled when the ball lands, consider the value of x when $y = -3.5$.

$$-3.5 = 2x - \frac{1}{2}x^2$$

$$x^2 - 4x - 7 = 0$$

$$x = 5.3166 \ldots \text{ or } x = -1.3166 \ldots$$

So: $x = 5.3$ m (1 d.p.)

(c) The horizontal velocity of the ball is $u\cos\theta$. In the horizontal, the velocity does not change as the ball moves along its trajectory, so it is the same when the ball reaches G.

Horizontal velocity $= u\cos\theta = \dfrac{7\sqrt{5}}{5} = 3.130 \ldots$

In the vertical, the initial velocity

$$= u\sin\theta = \frac{14\sqrt{5}}{5}.$$

Use the constant acceleration formulae to find the vertical velocity when the ball reaches G:

$$s = -3.5; u = \frac{14\sqrt{5}}{5}; v = ?; a = -9.8$$

$$v^2 = u^2 + 2as$$

$$v^2 = \left(\frac{14\sqrt{5}}{5}\right)^2 + 2(-9.8)(-3.5)$$

$$= 107.8$$

$$v = \sqrt{107.8} = 10.382 \ldots$$

Speed at G: $= \sqrt{3.130\ldots^2 + 10.382\ldots^2}$

$$= 10.8 \text{ m s}^{-1} \text{ (1 d.p.)}$$

Exercise 2C

In this exercise use $g = 9.8$ m s^{-2} unless you are instructed otherwise. Give your answers to a suitable level of accuracy.

1. A projectile is launched from ground level with an initial speed of 15 m s^{-1} at an angle of 20° to the horizontal. Find, giving your answers to one decimal place:
   (a) The projectile's maximum height.
   (b) The projectile's horizontal distance travelled.

2. A projectile is launched at an angle of 15° to the horizontal with a velocity of u m s^{-1}.
   Show that $u^2 = 2gR$
   where R is the horizontal range of the projectile.

3. (a) Write down the angle at which a projectile is launched to ensure the maximum range.
   (b) A volcano is erupting. As well as the lava and ash coming from it, there are numerous large rocks. Rocks are ejected with a maximum speed of 45 m s^{-1}. The town of Pericolo lies a horizontal distance of 2 km from the summit. Vertically, the volcano's summit is 500 metres above the town, as shown in the diagram.

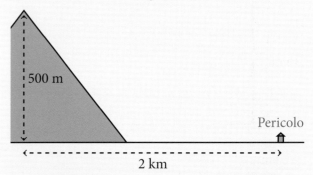

   Is it possible that Pericolo will be hit directly by one of the rocks coming from the volcano? Show all your working. You may use the equation for the trajectory of a projectile:

$$y = x\tan\theta - \frac{gx^2}{2u^2\cos^2\theta}$$

4. Tom hits a golf ball from horizontal ground with a speed of 30 m s^{-1} and at an angle of $\tan^{-1}\left(\dfrac{5}{12}\right)$. He hits the ball in the direction of a tree, which is at a horizontal distance of 16 m and has a height of 3 m. Show that the golf ball clears the tree, and calculate by how much. Give your answer in metres to 3 significant figures.

Exercise 2C...

5. On a battlefield a tank fires a shell with a velocity of $\sqrt{50g}$ m s^{-1} at an angle of 60° to the horizontal.
   (a) Find the range of the shell, giving your answer in simplified surd form.
   (b) Show that the equation of the shell's trajectory is
   $$y = -\frac{x^2}{25} + \sqrt{3}x$$
   Hence find the height of the shell when it has travelled horizontally $\sqrt{75}$ metres.

6. A stone is projected with speed 20 m s^{-1} from a point A on a cliff above horizontal ground. The point O on the ground is vertically below A and OA is 18 m. The stone is projected at an angle $\theta°$ to the horizontal. The point B is the highest point of the path of the stone and is 6 m above the level of A. The stone moves freely and hits the ground at the point C, as shown in the diagram.

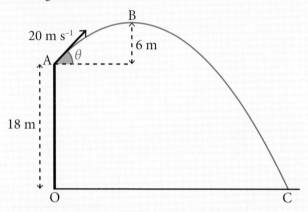

   Find:
   (a) The value of θ.
   (b) The distance OC.
   (c) The **speed** of the stone as it hits the ground at C.

7. A girl wants to throw a parcel onto a shelf. She stands 2 m away from the shelf horizontally. In the vertical, the shelf is 1.5 m above the parcel's initial position. She throws the parcel at 45°. At what speed should the girl throw the parcel for it to land on the shelf? Give your answer in the form $a\sqrt{bg}$, where a and b are integers to be found.

Exercise 2C...

8. A projectile leaves ground level with initial velocity u m s^{-1} and an angle of projection of 30°, as shown in the diagram.

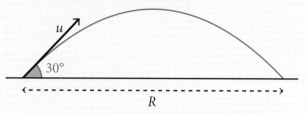

   After T seconds the vertical component of its velocity is 12 m s^{-1}. After $2T$ seconds the vertical component of its velocity is 7 m s^{-1}. Find the projectile's maximum height, giving your answer in terms of g.

9. A particle P is projected with a velocity of u m s^{-1} at an angle of $\theta°$ **below** the horizontal, where $\theta < 45°$, as shown in the diagram.

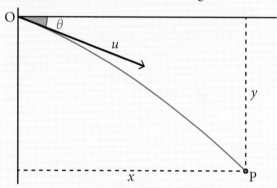

   (a) Show that the equation of the trajectory of this particle is:
   $$y = x\tan\theta + \frac{gx^2}{2u^2}\sec^2\theta$$
   (b) Find the time at which the particle has travelled the same distance in the vertical as it has in the horizontal. Give your answer in the form $t = \frac{ku}{g}f(\theta)$ where k is an integer constant to be found and $f(\theta)$ is an expression in terms of θ to be found.

10. At time $t = 0$, a rugby ball is kicked from a point O on horizontal ground with a velocity of u m s^{-1} and at an angle of $\theta°$ above the horizontal. When the ball is at point P, which is 3 m above the ground, it has a velocity of 12.9 m s^{-1} at an angle of 39° above the horizontal, as shown in the following diagram.

Exercise 2C...

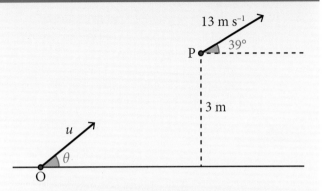

(a) Find u and θ, giving both answers to 3 significant figures.

(b) Hence find the time at which the ball is 3 m above the ground for the second time.

(c) State one modelling assumption you have made when answering this question.

11. Peter and Lucy are throwing stones against a rock on a beach. Peter says "The stone will lose energy when it hits the rock, so it will bounce back with a lower speed. Therefore, it can't bounce back this far and hit us." Is Peter right? Give a detailed explanation, referring to the maximum range of the stone as it returns.

2.3 Modelling Projectile Motion Using The Vector Forms Of The Constant Acceleration Formulae

In AS Mathematics you learned how to use the vector forms of the constant acceleration formulae. These three formulae are shown below.

$$\mathbf{v} = \mathbf{u} + \mathbf{a}t$$

$$\mathbf{s} = \mathbf{u}t + \frac{1}{2}\mathbf{a}t^2$$

$$\mathbf{s} = \frac{1}{2}t(\mathbf{u} + \mathbf{v})$$

You can use these vector forms to model an object moving in two dimensions under gravity.

In this section we will use $\mathbf{i}$ as the unit vector in the horizontal and $\mathbf{j}$ as the unit vector in the vertical.

A vector approach is used for the examples and exercise questions, but the scalar forms of the constant acceleration formulae could be used as a valid alternative approach. Unless a question specifically requires one of these approaches, it is acceptable to use either.

Worked Examples

8. A ball is thrown from an apartment balcony with an initial velocity of $(2\mathbf{i} + 3\mathbf{j})$ m s^{-1}. The ball takes 3 seconds to hit the ground. Take the acceleration due to gravity to be $-9.8\mathbf{j}$ m s^{-2} and give your answers to 1 decimal place where appropriate.

(a) Find the position of the ball when it hits the ground, relative to the balcony. Give your answer as a vector.

(b) Find the distance of the ball from the balcony at this time.

(c) State one modelling assumption that has been made.

(a) For the ball's trajectory for 3 seconds from the balcony:

$\mathbf{s} = ?$

$\mathbf{u} = 2\mathbf{i} + 3\mathbf{j}$

$\mathbf{v} =$

$\mathbf{a} = -9.8\mathbf{j}$

$t = 3$

The formula that doesn't involve $\mathbf{v}$ is:

$$\mathbf{s} = \mathbf{u}t + \frac{1}{2}\mathbf{a}t^2$$

$$\mathbf{s} = (2\mathbf{i} + 3\mathbf{j})(3) + \frac{1}{2}(-9.8\mathbf{j})(3)^2$$

$$= (6\mathbf{i} + 9\mathbf{j}) - 44.1\mathbf{j}$$

$$= (6\mathbf{i} - 35.1\mathbf{j}) \, \text{m}$$

(b) The ball is 6 metres horizontally and 35.1 metres vertically below its starting position. Find its distance using Pythagoras' Theorem.

$$\text{Distance} = |\mathbf{s}| = \sqrt{6^2 + 35.1^2}$$

$$= 35.6 \, \text{m (1 d.p.)}$$

(c) The ball is modelled as a particle.

9. At $t = 0$ seconds, an aircraft's position vector is $2000\mathbf{j}$ m relative to a fixed point O on the ground. At this time, a chunk of ice falls off the aircraft while it is travelling horizontally with velocity $200\mathbf{i}$ m s^{-1}.

(a) Find the **speed** of the ice after 10 seconds.

(b) A house lies under the flight path of the aircraft at position $3500\mathbf{i}$ m relative to O. Determine how far the ice lands from the house.

Give your answers to a suitable level of accuracy.

(a) For the first 10 seconds of the ice's journey:

$\mathbf{s} =$

$\mathbf{u} = 200\mathbf{i}$

$\mathbf{v} = ?$

$\mathbf{a} = -9.8\mathbf{j}$

$t = 10$

$\mathbf{v} = \mathbf{u} + \mathbf{a}t$

$\mathbf{v} = 200\mathbf{i} + (-9.8\mathbf{j})(10)$

$= 200\mathbf{i} - 98\mathbf{j}$

Speed $= |\mathbf{v}| = \sqrt{200^2 + 98^2} = 222.7196\ldots$

$= 220$ m s^{-1} (2 s.f.)

(b) On its journey to the ground, the ice travels 2000 metres downwards. We wish to calculate the horizontal distance x travelled.

For the ice's journey from the aircraft to the ground:

$\mathbf{s} = x\mathbf{i} - 2000\mathbf{j}$

$\mathbf{u} = 200\mathbf{i}$

$\mathbf{v} =$

$\mathbf{a} = -9.8\mathbf{j}$

$t = T$

$\mathbf{s} = \mathbf{u}t + \frac{1}{2}\mathbf{a}t^2$

$x\mathbf{i} - 2000\mathbf{j} = (200\mathbf{i})T + \frac{1}{2}(-9.8\mathbf{j})T^2$

Equating the **j**-components:

$-2000 = -4.9T^2$

$T^2 = \dfrac{2000}{4.9} = 408.163\ldots$

$T = 20.203\ldots$

Equating the **i**-components:

$x = 200T$

$= 200 \times 20.203\ldots$

$= 4040.61\ldots$

i.e. the ice lands at the point with position vector $4040.61\mathbf{i}$ m.

The house lies at $3500\mathbf{i}$ m, so the ice lands 540 m from it (2 s.f.)

Exercise 2D

Throughout this exercise:

- use a value of $g = 9.8$ m s^{-2} unless otherwise instructed.
- **i** and **j** are unit vectors in the horizontal and upwards vertical directions respectively.

Give your answers to an appropriate degree of accuracy.

2: PROJECTILES

Exercise 2D...

1. **In this question use a value of $g = 10$ m s^{-2}.**
   Initially a particle is at the point O with position vector $(0\mathbf{i} + 0\mathbf{j})$ m. It is projected with a velocity of $b\mathbf{i}$ m s^{-1}. After t seconds the particle is at the point with position vector $(15\mathbf{i} - 5\mathbf{j})$ m. Find the values of t and b.

2. In this question **i** and **j** are unit vectors in the horizontal and vertical direction respectively. A boy throws a tennis ball with initial velocity $(5\mathbf{i} + 1.5\mathbf{j})$ m s^{-1} from a point with a position vector of $0.5\mathbf{j}$ relative to a point O at ground level.
   (a) Find how long it takes for the ball to hit the ground.
   (b) State one modelling assumption that has been made.

3. Some sea birds fold their wings and dive into the water to catch fish. One such bird propels itself from a cliff, 44.1 metres high, with a velocity of $5\mathbf{i}$ m s^{-1}. It folds its wings and falls to the sea surface under the effects of gravity.
   (a) How long does it take the bird to reach the sea surface?
   (b) How far horizontally does it travel during its descent?
   (c) Find the bird's **speed** as it hits the water.
   (d) Why do the sea birds fold their wings when diving?

4. A lazy teacher gives the pupils their Applied Mathematics textbooks by throwing them from the front of the classroom. Damian's book travels with initial velocity $(7\mathbf{i} + 2.1\mathbf{j})$ m s^{-2}. Damian's desk has position vector $(4\mathbf{i} - 0.4\mathbf{j})$ m relative to the teacher.
   (a) How long does it take for Damian's book to travel 4 metres horizontally?
   (b) Determine whether Damian's textbook lands on his desk, goes too high, or too low. Show all your working.
   (c) What modelling assumptions have been used in your working? In what way may this affect your answer to part (b)?

5. Tilly fires a paintball pellet at Monty. She fires the pellet with a velocity of $(15\mathbf{i} + 0.5\mathbf{j})$ m s^{-1}. Monty is standing at a distance of 6 m horizontally from Tilly.

25

Exercise 2D...

(a) How long does it take the pellet to hit Monty?

(b) Monty is 1.7 metres tall. His legs are 80 cm long, his torso is 60 cm, his neck is 10 cm and his head 20 cm, as shown in the diagram.

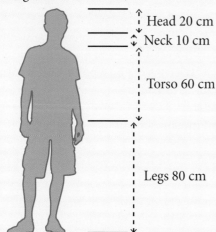

Head 20 cm

Neck 10 cm

Torso 60 cm

Legs 80 cm

Given that the pellet was fired from a height of 1.25 m, find where on Monty's body he is struck. Show all your working.

2.4 Summary

In this chapter you have learnt how to model the two-dimensional motion of objects moving as projectiles under the influence of gravity.

Section 2.2 demonstrated how this motion can be modelled using the scalar forms of the constant acceleration formulae in both the horizontal and vertical directions.

When the initial motion is horizontal, the vertical velocity is initially zero and this simplifies the equations. In these cases, use downwards as the positive direction, so that $g = 9.8$ m s^{-2}.

Section 2.3 demonstrated how this motion can be modelled using the vector forms of the constant acceleration formulae.

You will be required to know the formulae for range, maximum range, maximum height and time of flight for a projectile, along with the general equation of the trajectory. You are also required to learn the proofs for these five formulae.

Chapter 3
Moments

3.1 Introduction

In AS Applied Mathematics you learnt that a **particle** is in equilibrium if there is no resultant force acting upon it. A particle is a theoretical object that has a mass, but no size.

In this chapter, objects are modelled as **light rods**. A **rod** is a theoretical object with a length, but no width or height, i.e. it is one-dimensional. A rod is **light** if its mass can be neglected.

This chapter provides an introduction to **moments**. A moment is the turning effect of a force. For a rod to be in equilibrium, there must be no resultant force **and** the sum of the **moments** acting must be zero.

The following two chapters build on this work by considering a rod's mass. Beams, seesaws and ladders are modelled as rods and exam standard questions are presented.

Key words
- **Particle**: An object with a mass but no size.
- **Rod**: A one-dimensional rigid body. It has a length, but no width or height. The mass of a rod is concentrated along a line.
- **Rigid body**: A one, two or three-dimensional object that does not bend or break. A rod is a type of rigid body.
- **Light**: An object is said to be light if its mass can be neglected.
- **Equilibrium**: An object is in equilibrium if it is not moving and not turning.
- **Moment**: The turning effect of a force about a point.
- **Coplanar**: Coplanar forces are forces that act in the same plane. Only coplanar forces are considered in this chapter and the next two.

Before you start
You should know:
- The definitions of the words **particle**, **rod** and **light**.
- Basic trigonometry.
- How to resolve a force into two components.

Exercise 3A (Revision)

1. (a) In mechanics, what is meant if something is modelled as:
   (i) a particle?
   (ii) a rod?
   (b) What is meant by a 'light rod'?

2. Find the missing sides in these triangles.
   (a)

   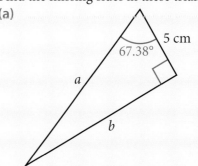

   (b)

   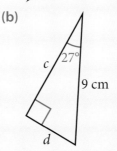

3. The particle P shown experiences a force of 35 N at an angle of 27° to the horizontal. Resolve the force into horizontal and vertical components.

   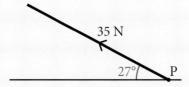

4. The diagram shows a box of mass 2 kg on a smooth plane inclined at 20° to the horizontal.

   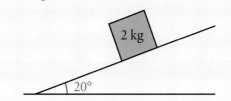

Exercise 3A...

(a) Copy the diagram and add all the forces acting on the box.

(b) Resolve the box's weight force into two components, parallel and perpendicular to the slope.

What you will learn

In this chapter you will learn:

- The concept of a moment as the turning effect of a force.
- About the sense of a moment.
- About the sum of moments.
- About rods in equilibrium.
- About the moment from oblique forces.

In the real world...

In 1966, before Neil Armstrong became famous as the first person to walk on the moon, he was Command Pilot on Gemini 8, alongside fellow astronaut Dave Scott. This was the first attempt to dock, or connect, two spacecraft, a vital part of the preparation for the later missions to the moon. The docking procedure went well and Scott radioed back to mission control 'Flight, we are docked! Yes, it's really a smoothie!'

Shortly after docking, however, the combined craft started to turn, or roll. The astronauts undocked, but their craft continued to roll, faster and faster, so that the spacecraft and astronauts were now rotating at almost one revolution per second. Without their gruelling physical training, the astronauts would have passed out. Armstrong fired the thrusters to attempt to correct the problem, but it kept recurring. Eventually, low on fuel, he shut down the thruster system and activated the re-entry thrusters, which were supposed to guide the spacecraft back to Earth. With these he was able to stop the spin. He also saved enough fuel for a successful return through the Earth's atmosphere. Later investigations discovered one of the thrusters had been stuck on, providing a continuous force that caused the spacecraft to spin.

Neil Armstrong had shown he could work well under extreme pressure. His quick thinking and expertise meant he was later chosen to command the Apollo 11 mission to the moon, and to become the first person to set foot on the lunar surface.

3.2 The Moment Of A Force

To describe a force acting on a **particle** you need to know the magnitude and direction of the force. The effect of the force on the particle is to produce a **translation**: the particle moves in a straight line without rotating.

When a force is applied to a **rigid body** the point of application of the force must be considered as well as its magnitude and direction. The force may cause the body to rotate, as well as experience the translational motion.

The turning effect of a force about a point depends on both the magnitude of the force and its distance from the point.

The **moment** of a force is a measure of the force's ability to rotate the object on which it acts. It is the product of the magnitude of the force and the perpendicular distance from the point to the line of action of the force.

Sense of the rotation

Consider a man sitting on one end of a seesaw, as shown below.

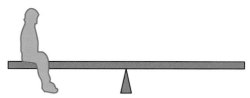

The man's weight force causes the seesaw to turn anticlockwise. This is the **sense of the rotation**.

If the man sat on the other end, the sense of the rotation would be clockwise.

Use a clockwise arrow such as ↷ to denote the clockwise direction and an anticlockwise arrow such as ↶ to denote the anticlockwise direction.

Note: When resolving forces, refer to the direction in which forces are being resolved, e.g. horizontal, vertical, perpendicular to the rod, etc.

When taking moments, always refer to the point about which moments are being taken, as in the following example.

Worked Example

1. In the following diagrams, a single force acts at a point on a light rod. In each case, find the moment of the force about the fixed point C.

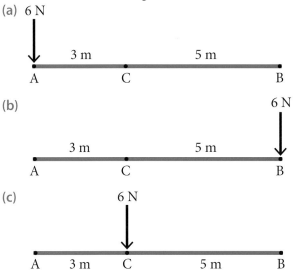

(a) 6 N

(b) 6 N

(c) 6 N

Note: Since, in this example, the rod is light, its weight force can be neglected.

(a) The size of the moment of the 6 N force **about the fixed point C** is:

Force × perpendicular distance
= 6 × 3
= 18 N m

With the force acting downwards at A, the rod starts to rotate in an anticlockwise direction about the fixed point C.

Therefore, the moment is 18 N m anticlockwise.

Note: The unit for moments is newton metres (N m), since a moment is the product of a force and a distance.

(b) The size of the moment of the force **about point C** is:

Force × perpendicular distance
= 6 × 5
= 30 N m clockwise

(c) The size of the moment of the force **about point C** is:

Force × perpendicular distance
= 6 × 0
= 0 N m

Note: If the line of a force passes through a point, the force provides a moment of zero newton metres about that point.

The resultant moment

The **resultant moment** is the sum of the moments.

When several forces act on a body, the resultant moment about a point can be found by adding the moments from all the forces, taking the sense of rotation for each one into account.

Worked Example

2. Three coplanar forces of magnitude 3 N, 4 N and 5 N act upon a light rod AC of length 6 m, as shown in the diagram. Calculate the resultant moment of these forces about A.

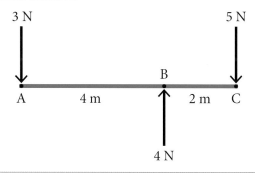

3 N   5 N

A   4 m   B   2 m   C

4 N

The moment of the 3 N force about A is
$3 \times 0 = 0$ N m

The moment of the 4 N force about A is
$4 \times 4 = 16$ N m anticlockwise.

The moment of the 5 N force about A is
$5 \times 6 = 30$ N m clockwise.

As the clockwise moment is greater than the anticlockwise moment, calculate the sum of the moments in the clockwise direction:

The resultant moment about point A is
$30 - 16 = 14$ N m clockwise.

Equilibrium

In AS Mathematics you learnt about **particles** in equilibrium. For a particle to be in equilibrium, forces acting upwards must be equal to those acting downwards, and forces acting to the left must be equal to those acting to the right. That is, there is no **resultant** force acting on the particle.

For a **rigid body** to remain in equilibrium:

• there is no resultant force; and

• the resultant moment is zero. Expressed another way, the anticlockwise moments are equal to the clockwise moments.

As in the following example, you may be asked to find the resultant moment acting on a rod. In this way you can determine whether the rod is in equilibrium, or whether a rotation takes place.

....

Worked Example

3. The light rod AB in the diagram has length 2 m and a fixed centre O. Coplanar forces of size 2 N act at points A and B, as shown. Is the rod in equilibrium? Give a reason for your answer.

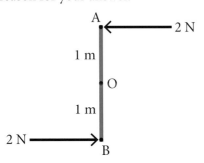

For the rod to be in equilibrium, there must be no resultant force **and** the resultant moment must be zero.

The resultant force is zero both vertically and horizontally.

Take moments about the fixed centre O.

Consider the force acting at point A. The size of the force is 2 N and the perpendicular distance to O is 1 m. Therefore, there is an anticlockwise moment of size $2 \times 1 = 2$ N m.

The force acting at B also provides an anticlockwise moment of size 2 N m.

Overall, the sum of the moments is 4 N m anticlockwise. Therefore, the rod is not in equilibrium; it would rotate about its fixed centre O in an anticlockwise direction.

....

If a rod is in equilibrium, anticlockwise moments are equal to clockwise moments.

In addition, it is possible to resolve forces, usually perpendicular to the rod.

By considering both of these properties of equilibrium, it is possible to find two unknown forces.

....

Worked Example

4. Three coplanar forces act upon the light rod shown in the diagram. The rod is in equilibrium. Find the forces F_1 and F_2.

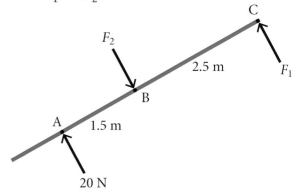

Take moments about point B. Since the rod is in equilibrium:

anticlockwise moments = clockwise moments

$$F_1 \times 2.5 = 20 \times 1.5$$
$$2.5F_1 = 30$$
$$F_1 = 12 \text{ N}$$

> **Note:** By taking moments about point B, F_2 does not appear in the equation. In this way it is possible to find an equation in one unknown, F_1, which can be solved.

Resolve forces perpendicular to the rod. Since the rod is in equilibrium:

$$F_2 = F_1 + 20$$
$$F_2 = 32 \text{ N}$$

....

Exercise 3B

In the following questions all forces are coplanar and acting either parallel or perpendicular to the rod.

1. Find the resultant moment about point O provided by the forces acting on the light rods shown. Also state the sense of this resultant moment.

    (a)

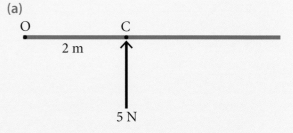

Exercise 3B...

(b)

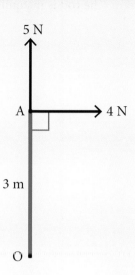

5 N

A ⟶ 4 N

3 m

O

(c)

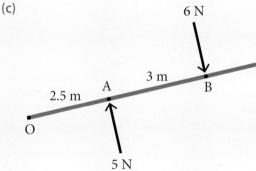

6 N

3 m

A   B

2.5 m

O

5 N

(d)

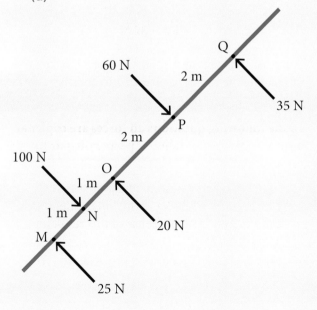

60 N

Q

2 m

35 N

P

2 m

100 N

O

1 m

N

1 m

20 N

M

25 N

Exercise 3B...

2. For each of the light rods shown below, determine whether it is in equilibrium. Show all your working.

(a)

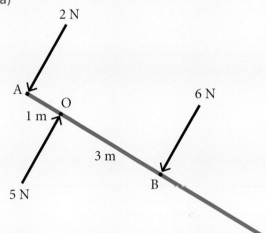

2 N

A

1 m

O

6 N

3 m

B

5 N

(b)

100 N ⟶ C ⟵ 160 N

3 m

O

1 m

60 N ⟶ A

(c)

250 N

P   2 m   3 m   R

Q

150 N   100 N

Exercise 3B...

3. The light rods shown below are in equilibrium. Find the missing forces in each diagram.

(a)

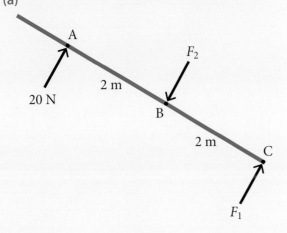

(b)

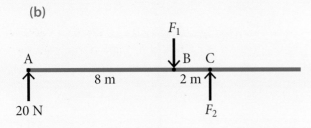

(c)

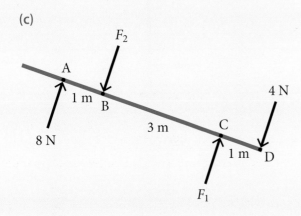

Exercise 3B...

(d)

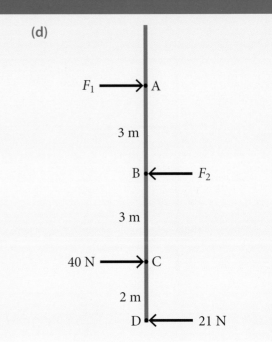

3.3 The Moment From An Oblique Force

A force that does not act parallel or perpendicular to a rod can be referred to as an **oblique force**.

In the following example, a force acts obliquely upon a rod. For the calculation of the moment provided by this force, two methods are given.

....

Worked Example

5. The diagram shows a light rod AB, length 5 m. A force of 10 N acts at B at an angle of 37° to the rod. Calculate the moment provided by this force about point A.

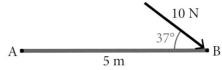

The force is not acting at right angles to the rod. In this case the moment is calculated using one of the two methods below.

In both methods, the force is shown as an arrow coming out of the point B.

Method 1

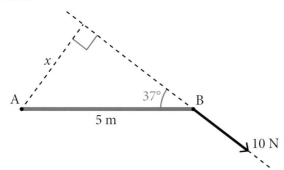

Calculate the perpendicular distance between the line of the force and the centre of rotation. This perpendicular distance is marked x in the diagram.

From the right-angled triangle, it can be seen that

$$\sin 37 = \frac{x}{5}$$

$$\Rightarrow x = 5 \sin 37 = 3.01 \, \text{m}$$

The size of the moment provided by the 10 N force is:

Force × perpendicular distance
$$= 10 \times 3.01$$
$$= 30.1 \, \text{N m (3 s.f.)}$$

Method 2

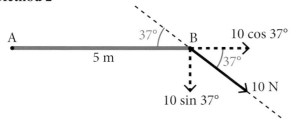

Resolve the 10 N force into two components, parallel and perpendicular to the rod, shown as dashed arrows in the diagram. The component perpendicular to the rod (acting in the vertical) is $10 \sin 37$

The perpendicular distance from the line of this force to point A is 5 m. Therefore, the moment provided is

$$10 \sin 37 \times 5 = 30.1 \, \text{N m (3 s.f.)}$$

> **Note:** The component parallel to the rod can be ignored. The moment provided by this component is zero, since the line of this force passes through point A.

More generally, if a force of magnitude F N acts obliquely on a rod at an angle of $\theta°$ and a distance of d metres from a point, the moment provided by this force about the point is $Fd \sin \theta$ N m.

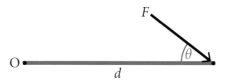

The moment about O from the oblique force of magnitude F N is $Fd \sin \theta$ N m clockwise.

Equilibrium

In the previous section, the equilibrium of rods was discussed. For a rigid body, such as a rod, to remain in equilibrium, there must be no resultant force and no resultant moment.

In this section, the idea is extended so that oblique forces are considered.

In the following example, two forces act obliquely upon a rod, along with two other forces acting perpendicular to the rod.

....

Worked Example

6. The light rod shown is acted upon by four coplanar forces, as shown. Given that the rod is in equilibrium, find the magnitude of forces F_1 and F_2.

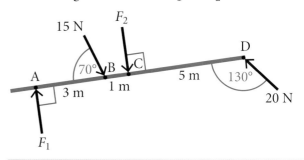

Take moments about point A.

The moment provided by an oblique force can be calculated as $Fd \sin \theta$ N m.

So the 15 N force provides a clockwise moment of $15 \times 3 \times \sin 70 = 45 \sin 70$ N m.

The 20 N force provides an anticlockwise moment of $20 \times 9 \times \sin 130 = 180 \sin 130$ N m.

Since the rod is in equilibrium:

clockwise moments = anticlockwise moments:

$$45 \sin 70 + 4F_2 = 180 \sin 130$$
$$42.286 + 4F_2 = 137.888$$
$$4F_2 = 95.602$$
$$F_2 = 23.900 = 23.9 \, \text{N (3 s.f.)}$$

Resolve forces perpendicular to the rod. The diagram has been redrawn, showing only the perpendicular components of the forces.

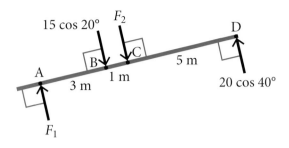

$$15 \cos 20 + F_2 = F_1 + 20 \cos 40$$

$$14.095 + 23.900 = F_1 + 15.321$$

$$F_1 = 22.674 = 22.7 \, \text{N (3 s.f.)}$$

Exercise 3C

In the following questions all forces can be assumed to be coplanar.

1. For each of the light rods that follow, determine whether the rod is in equilibrium. Show all your working.

   (a)

   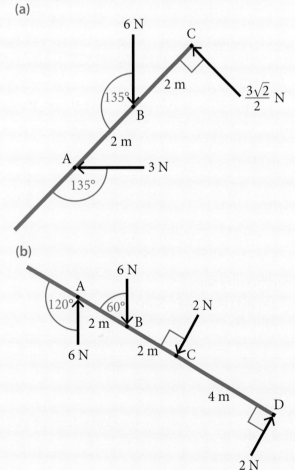

   (b)

Exercise 3C...

2. Find the resultant moment about point O provided by the forces acting on each of the light rods shown. Also state the sense of this resultant moment.

   (a)

   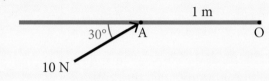

   (b)

   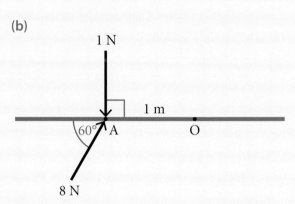

   (c)

   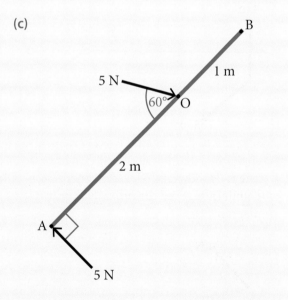

Exercise 3C...

(d)

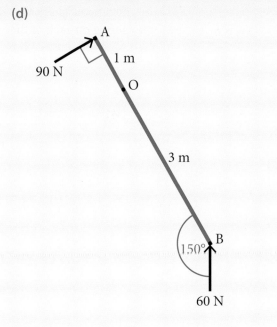

(e)

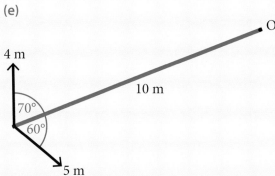

3. The rods shown below are light and in equilibrium. Find the missing forces in each diagram.

(a)

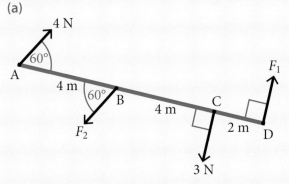

Exercise 3C...

(b)

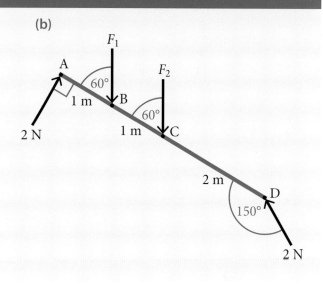

3.4 Summary

This chapter has considered forces acting upon **light rods**. A rod is a theoretical rigid body with a length, but no width or height, i.e. it is one-dimensional. A rod is light if its mass is negligible.

For a force acting on a rod, the **moment** about a point O is the product of the magnitude of the force and the perpendicular distance from the line of the force to the point O.

For a force acting perpendicular to the rod, as shown in the following diagram:

Moment $= Fd$ N m

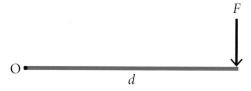

For a force acting obliquely to the rod, as shown in the following diagram:

Moment $= Fd \sin \theta$ N m

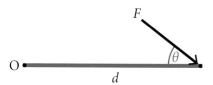

For a force measured in newtons and a distance in metres, the moment is measured in newton metres (N m).

The **sense** of a moment can be clockwise or anticlockwise.

The sum of the moments acting on a body is called the **resultant moment**.

For a rod in equilibrium:

- There is no resultant force; and
- There is no resultant moment, i.e. clockwise and anticlockwise moments are equal.

Chapter 4
Horizontal Rods

4.1 Introduction

In the previous chapter you learnt about forces acting on **rods**. You learnt that a force can have a turning effect called a **moment**. A rod is in equilibrium if there is no resultant force and no resultant moment.

This chapter considers objects modelled as horizontal **rods** in equilibrium. In mechanics, a **rod** is a one-dimensional rigid body. A rod may have a mass, or it may be light, in which case its mass is neglected. See-saws and horizontal beams are examples of objects that may be modelled as rods.

Throughout this chapter, it may be assumed that all forces mentioned in a question or example are **coplanar**, i.e. that the lines of all the forces are within the same plane.

Key words

- **Particle**: An object with a mass but no size.
- **Rigid body**: A one-, two- or three-dimensional object that does not bend or break.
- **Rod**: A one-dimensional rigid body. It has a length, but no width or height. The mass of a rod is concentrated along a line.
- **Equilibrium**: An object is in equilibrium if it is not moving and not turning.
- **Moment**: The turning effect of a force about a point.
- **Coplanar**: Coplanar forces are forces that act in the same plane.
- **Uniform**: The mass of a uniform body is evenly distributed along its length. Its weight force acts at the midpoint.
- **Sharp**: If a rod rests on a sharp support, that support touches the rod at a single point.
- **Inextensible**: A string that is inextensible does not stretch.
- **Light**: A string or rod that is light has a negligible mass.

Before you start

You should know:

- That a moment is the turning effect of a force.
- How to calculate a moment for a force acting perpendicular to a rod.
- How to calculate a moment for a force acting obliquely to a rod.
- That in equilibrium there is no resultant force and no resultant moment.
- That a moment is either clockwise or anticlockwise. This is known as the **sense** of the moment.

Exercise 4A (Revision)

1. Explain briefly the two conditions required for a rigid body to be in equilibrium.

2. Calculate the resultant moment about point O from the system of two forces acting upon the light rod shown in the diagram below.

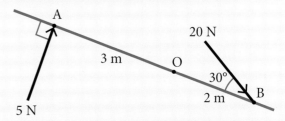

3. The rod AB shown in the diagram is in equilibrium. Find the magnitudes of forces F_1 and F_2.

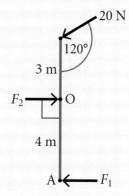

What you will learn

In this chapter you will learn:

- How to use moments in simple static contexts, such as horizontal beams resting on pivots or supported by strings.
- How to use moments with non-uniform rods.
- How to use moments with horizontal rods on the point of tilting or turning.

In the real world...

The Second Severn Crossing, or the Prince of Wales Bridge, is the bridge that takes the M4 motorway across the River Severn between England and Wales. It carries thousands of cars every day.

The bridge is a cable-stayed bridge; in this type of bridge steel cables attached to one or more towers support the platform. In the case of the Second Severn Crossing, there are two towers.

Is each section of the bridge strong enough to stand up to the volume of traffic it could experience? Engineers answer this question using computer models that analyse the forces and moments that the structure will experience.

4.2 Finding The Magnitude Of Reaction Forces

When a horizontal rod rests upon supports, each support exerts an upwards **reaction force** upon the rod. In this section moments are used to find the size of one or more of these reaction forces.

Note: A common strategy is to take moments about one of the points of support. In this way it is possible to find an equation in one unknown, which can be solved to find one reaction force. Then resolve forces in the vertical to find the second reaction force.

Worked Example

1. The uniform plank AB shown has a mass of 60 kg. It rests in equilibrium on the supports at points P and Q as shown.

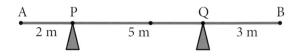

(a) Copy the diagram, marking all the forces acting upon the plank.

(b) Find the sizes of the reaction forces at the points P and Q.

(c) What modelling assumptions have been made?

(a) Since the plank is uniform, the weight force acts in the centre. This point lies 5 metres from each end. It is 3 m from P.

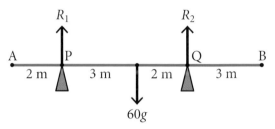

(b) Take moments about P. In this way, the force R_1 does not appear in the equation and R_2 can be found:

$$60g \times 3 = R_2 \times 5$$
$$5R_2 = 180g$$
$$R_2 = 36g = 353 \, \text{N}$$

Resolving forces in the vertical:

$$R_1 + R_2 = 60g$$
$$R_1 + 36g = 60g$$
$$R_1 = 24g = 235 \, \text{N}$$

(c) The plank is modelled as a rod. The supports are assumed to be sharp.

The following example involves a light beam. Notice that a diagram has not been given and is not asked for. However, it is important to draw one to gain a full understanding of the situation.

Worked Example

2. A light beam AB of length 5 m rests on horizontal supports at C and D, where AC = 1 m and BD = 1 m. A mass of 10 kg rests on the beam, 2 m from A. Find the reaction forces at both supports, giving your answers in terms of g.

The beam is light, so the only weight force downwards upon the beam is from the additional mass.

All distances have been calculated and added to a diagram.

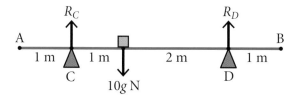

Taking moments about point C:

$$10g \times 1 = R_D \times 3$$

$$R_D = \frac{10g}{3} \text{ N}$$

Resolving forces in the vertical:

$$R_C + R_D = 10g$$

$$R_C = 10g - \frac{10g}{3} = \frac{20g}{3} \text{ N}$$

The following example involves both the weight force of the beam and an object on the beam, providing an additional downwards force.

Worked Example

3. A uniform beam AB of length 6 m and mass 40 kg rests horizontally on supports at C and D, where AC = BD = 1.2 m, as shown in the diagram. A box of mass 15 kg rests on the beam at E, where AE = 2.5 m.

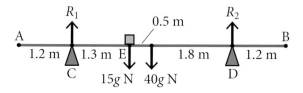

(a) Copy the diagram, adding all the forces acting upon the beam.

(b) Calculate the magnitude of the reaction force at each of the supports C and D.

(a) Let the reaction at C be R_1 N and the reaction at D be R_2 N. The distances between points are also shown.

(b) Take moments about point C. Since the line of force R_1 passes through this point, R_1 does not appear in the equation.

$$(15g \times 1.3) + (40g \times 1.8) = R_2 \times 3.6$$

$$19.5g + 72g = 3.6R_2$$

$$91.5g = 3.6R_2$$

$$R_2 = 249.083 \dots$$

$$R_2 = 249 \text{ N (3 s.f.)}$$

Since the beam is in equilibrium, forces acting upwards upon it are equal to those acting downwards.

$$R_1 + R_2 = 15g + 40g$$

$$R_1 + 249.083 = 55g$$

$$R_1 = 289.916 \dots$$

$$R_1 = 290 \text{ N (3 s.f)}$$

Exercise 4B

1. A light beam AB of length 1 m rests on horizontal supports at A and B. A mass of 1 kg rests on the beam, 0.75 m from A. Find the reaction forces at both supports, giving your answers in terms of g.

2. A uniform beam AB of mass 10 kg and length 5 m rests on two supports at C and D, where AC = 1 m and BD = 2 m.
    (a) Draw a diagram showing all the forces acting on the beam.
    (b) Find the magnitude of the reaction forces at C and D, giving your answers in terms of g.

3. The diagram shows a uniform plank AB of length 2.5 m and mass 20 kg. The plank rests horizontally in equilibrium upon supports at A and C, where CB = 0.5 m. A toolbox of mass 8 kg is placed at point T on the plank between the two supports, such that TC = 0.3 m.

    (a) Copy the diagram, adding all external forces acting upon the plank.
    (b) Find the magnitude of the reaction forces at both supports.
    (c) State any modelling assumptions made in answering this question.

4. Two workmen, Chris and Eamonn, of mass 57 kg and 75 kg respectively, stand on a uniform plank ABCDEF of length 5 m and mass 50 kg. The plank is supported at B and D, where B is 1 m from A and D is 1 m from F. Chris stands at C, which is 1.5 m from A and Eamonn stands at E which is 0.5 m from F.
    (a) Draw a diagram showing all the forces acting upon the plank.

Exercise 4B...

(b) Find the magnitude of the reaction force at each support, giving your answers in newtons to 3 significant figures.

(c) State one modelling assumption that has been made.

5. A gymnast of mass 40 kg performs on a uniform beam AB, which is 5 metres long and has a mass of 50 kg. The beam is supported at points C and D, which are 50 cm from A and B respectively. As the gymnast begins her routine, she stands stationary at point G, such that the distance GB is four times the distance AG.

    (a) By considering the forces acting upon the beam, find the magnitude of the reaction forces at the two supports at this time.

    (b) State two modelling assumptions that have been made.

4.3 Finding The Position Of A Force

Consider two forces of unequal magnitude, for example the weight forces of a man and a boy on a see-saw, as in the next example. For them to have an equal turning effect about the pivot, the man must sit closer to the pivot. In this way the see-saw remains in equilibrium.

Worked Example

4. A man of mass 60 kg sits on the left of a see-saw, while his son of mass 40 kg sits on the right, as shown.

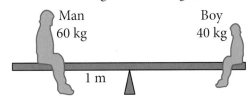

The man sits 1 metre from the pivot. How far must the boy sit for the see-saw to remain in equilibrium?

The moment provided by the man's weight force is the product of his weight force and the distance to the pivot, i.e. ⸰

Man's moment $= 60g \times 1 = 60g$ N m anticlockwise

For the see-saw to remain in equilibrium, the boy must provide a clockwise moment of $60g$. If his distance from the pivot is x, then:

$40g \times x = 60g$

$\qquad x = 1.5\,\text{m}$

In the following example we are asked to find the position of one of the supports for a rod in equilibrium. Again, take moments about a point on the rod.

Worked Example

5. The diagram below shows a uniform rod AB of length 1 m, supported at points C and D, where AC = 0.3 m. The rod rests horizontally in equilibrium. The magnitude of the reaction force at point C is double the magnitude of the reaction force at point D.

Find the position of point D, giving your answer in centimetres from A.

The diagram should be redrawn with all the forces acting upon the rod added.

If the reaction at D is R N, then the reaction at C is $2R$ N.

Let the mass of the rod be M kg.

Let the distance between point D and the centre of the rod be x metres.

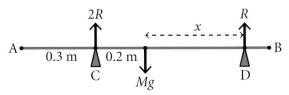

Take moments about the centre of the rod. Since this is a uniform rod, the line of the weight force passes through the centre. Therefore Mg does not appear in the equation.

$2R \times 0.2 = R \times x$

$\qquad 0.4R = Rx$

$\qquad\quad x = 0.4\,\text{m}$

The question requires the distance AD.

$\qquad AD = 0.3 + 0.2 + 0.4 = 0.9\,\text{m}$

Point D is 90 cm from point A.

Exercise 4C

1. Conor and his younger sister Emily are on either side of a uniform see-saw, which is smoothly pivoted in the centre. Conor has a mass of 30 kg and sits 90 cm from the pivot. Emily's mass is 18 kg. Find how far from the pivot she should sit for the see-saw to remain horizontally balanced.

2. A tortoise of mass 3.4 kg walks up a ramp at a steady speed of 0.05 m s^{-1}. The ramp can be modelled as a uniform rod with midpoint M and length 3 m. It is smoothly pivoted at M. Sitting on the ramp, on the other side of the pivot, is a cat of mass 5.1 kg, as shown in the diagram. If the tortoise starts walking from the end of the ramp, find how long it takes for the ramp to become horizontal.

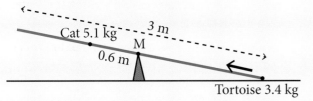

3. The diagram below shows a uniform plank AB of length 2 m, supported at points P and Q, where AP = 0.3 m. The plank rests horizontally in equilibrium. The magnitude of the reaction force at point Q is double the magnitude of the reaction force at point P.

   (a) Copy the diagram, adding all the forces acting upon the plank.
   (b) Find the distance AQ.
   (c) State one modelling assumption that has been made in answering this question.

4. A pencil has length 10 cm and mass 10 grams. Its mass is distributed uniformly along its length. It rests on two pencil sharpeners, the first of which is 2 cm from the pencil's point. The reaction force provided by this sharpener is 1.5 times the reaction force provided by the second sharpener.
   (a) Find the reaction force provided by each sharpener.
   (b) Find the distance between the point of the pencil and the second sharpener.

Exercise 4C...

   (c) State any modelling assumptions that have been made in answering parts (a) and (b).

5. Gary and Eimar and their daughter Mia sit on a park bench, with Gary on the left.
   (a) Find how far from the left-hand end of the bench Gary is sitting, giving your answer to the nearest centimetre.
   You may use the following information in your calculations:
   - The bench has a length of 2 metres.
   - The bench is supported by two legs, which are positioned directly beneath the two ends of the bench.
   - Excluding the legs, the mass of the bench is 150 kg.
   - The bench's mass is distributed uniformly along its length.
   - The reaction forces in the two legs are 160g N and 145g N respectively.
   - Gary, Eimar and Mia have masses of 70 kg, 55 kg and 30 kg respectively.
   - Eimar's centre of mass is 150 cm from the left-hand end.
   - Mia sits exactly halfway between her parents.
   (b) What modelling assumptions have you made in answering part (a)?

4.4 Finding The Mass Of An Object

You may be asked to find the mass of a rod or the mass of an additional object upon it.

The following example demonstrates finding the mass of a beam using moments.

....

Worked Example

6. A uniform horizontal beam AB of length 8 m and mass M kg rests upon two supports at C and D where AC = 2 m and BD = 1.5 m. The reaction force at support C is 30 N. Find:
   (a) M; and
   (b) the size of the reaction force at support D.

   Draw a diagram, adding any missing lengths.

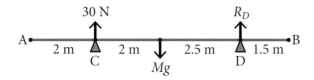

(a) Take moments about point D. In this way, R_D does not appear in the equation.

$$Mg \times 2.5 = 30 \times 4.5$$
$$2.5Mg = 135$$
$$M = \frac{54}{g} = 5.51 \text{ kg (3 s.f.)}$$

(b) Resolve forces in the vertical.

Since $M = \frac{54}{g}$, $Mg = 54$:

$$30 + R_D = Mg$$
$$30 + R_D = 54$$
$$R_D = 24 \text{ N}$$

Exercise 4D

1. Peadar and Roisin sit on a see-saw, which is modelled as a uniform rod smoothly pivoted at its centre of mass. Peadar has a mass of 30 kg and sits 1 m from the pivot on the left. Roisin is sitting 1.2 m from the pivot on the right. Given that the see-saw is horizontal and in equilibrium, find Roisin's mass in kilograms.

2. A uniform horizontal plank AB of length 10 m and mass M kg rests upon two supports at P and Q where AP = 2.5 m and BQ = 3 m. The reaction force at support P is 120 N.
   (a) Draw a diagram, showing all the forces acting upon the plank.
   (b) Find:
      (i) M;
      (ii) the size of the reaction force at support Q.
   (c) State any modelling assumptions used.

3. A uniform rod AB has length 2 m and mass M kg. It is supported at points C and D, where AC = 0.6 m. The reaction forces at the supports at C and D are R_C and R_D respectively.
   (a) Given that $R_D = R$ and $R_C = 2R$, find M in terms of R and g.
   (b) Find the distance BD.
   (c) An additional mass of m kg is now placed on the rod at A and the rod remains in

Exercise 4D...

equilibrium. The reaction force at C is now 5 times the reaction force at D. Find m in terms of M.

4. A car of mass M kg drives across a bridge. The roadway of the bridge has a mass of 2000 kg, which is uniformly distributed. The roadway is supported by two concrete pillars, P_1 and P_2, which are positioned 4 metres either side of the centre of mass, with P_1 on the left. The car passes above P_1 and the reaction forces at P_1 and P_2 are R_1 and R_2 newtons respectively.
   (a) Find R_2
   (b) Given that $R_1 : R_2 = 5 : 2$, find M.
   (c) State one modelling assumption that has been made about each of:
      (i) the car;
      (ii) the roadway;
      (iii) the concrete pillars.

5. A road bridge across a river is 500 m long and has a mass of 10^5 kg. It can be modelled as a single uniform beam resting on two supports A and B, as shown in the diagram. Each support is 100 m from one end of the bridge.

The reaction force at either support should not exceed 10^8 newtons. If either reaction force exceeds this, the bridge is considered unsafe. Point M is the midpoint of the bridge and point E lies halfway between M and B.
   (a) Estimate the maximum number of vehicles that could be on the bridge at any time. You may assume there are two lanes of traffic **in each direction** and that the average length of a vehicle is 1 m.
   (b) A traffic jam builds up between the centre of the bridge and support B. This can be modelled as a single particle at point E. The jam comprises 50 vehicles, each with a mean mass of 1500 kg.
      (i) Find the total mass of this traffic jam.
      (ii) Assuming there are no other vehicles on the bridge, copy the diagram, adding all the forces acting upon the road bridge.
      (iii) Find the reaction forces at the two supports and determine whether the bridge is safe.

Exercise 4D...

(c) The traffic at point E clears. Later in the day a new traffic jam builds up around support B.

   (i) How could this traffic jam be modelled?

   (ii) Again assuming that there are no other vehicles on the bridge, copy the diagram, adding all the forces now acting upon the road bridge.

   (iii) Find the reaction force at support A.

   (iv) What mass of traffic in this jam would cause the bridge to become unsafe? Discuss briefly whether this is ever likely to occur, stating any assumptions you have made in your calculations.

4.5 Rods Supported By Strings

A beam may be suspended horizontally by strings. It experiences upwards forces due to the tension in each string.

In the following example, a distance must be calculated. The method is essentially the same as described in the previous section.

....

Worked Example

7. A uniform beam AB of mass 8 kg and length 4 metres is suspended horizontally from a ceiling by two strings attached at points C and D, as shown in the diagram. AC = 0.5 m and DB = 1 m. A box of mass 2 kg rests on the beam at its centre of mass.

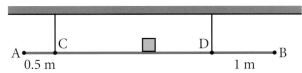

(a) Copy the diagram, adding all the forces acting upon the beam.

(b) Find the magnitude of the tension in both strings.

(c) How far towards point B should the box be moved so that the tension in the string at D becomes twice the tension in the string at C?

(a) The weight force acting at the centre of the beam is the sum of the weights of the beam and the box, that is $8g + 2g = 10g$.

Let T_1 and T_2 be the tensions in the strings at C and D respectively.

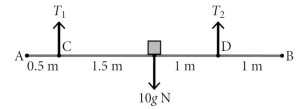

(b) Take moments about C:

$$10g \times 1.5 = T_2 \times 2.5$$
$$2.5T_2 = 15g$$
$$T_2 = 6g \text{ N}$$

Since the beam is in equilibrium, the sum of the forces upwards is equal to the sum of the forces downwards:

$$T_1 + T_2 = 10g$$
$$T_1 + 6g = 10g$$
$$T_1 = 4g \text{ N}$$

(c) Let the distance the box is moved to the right be x metres. Since the tension in the string at D is twice the tension in the string at C, label them $2T$ and T respectively.

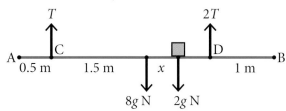

Since the beam is still in equilibrium in the vertical,

$$T + 2T = 8g + 2g$$
$$3T = 10g$$
$$T = \frac{10}{3}g$$

Taking moments about the centre:

$$(T \times 1.5) + (2g \times x) = 2T \times 1$$
$$1.5T + 2gx = 2T$$
$$2gx = 0.5T$$
$$x = 0.25\frac{T}{g}$$

Since $T = \frac{10}{3}g$,

$$x = 0.25 \times \frac{10}{3} = \frac{5}{6} \text{ metres, or } 83.3 \text{ cm (3 s.f.)}$$

....

Exercise 4E

1. A light beam AB of length 1 metre is suspended horizontally by two ropes attached to the ends A and B. A box of mass 1 kg is placed on the beam. How far from A should the box be placed so that the tension in the rope at A is double the tension in the rope at B? Give your answer in centimetres.

2. A uniform beam AB of mass 10 kg and length 2 metres is suspended horizontally from a ceiling by two light, inextensible strings attached at points C and D, as shown in the diagram. AC = 0.5 m and DB = 0.2 m. Find the tension in each string.

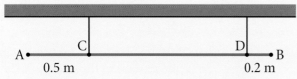

3. A uniform beam AB of mass 20 kg and length 5 metres is suspended horizontally from a ceiling by two light, inextensible strings attached at points C and D, as shown in the diagram. AC = 0.8 m and DB = 1.5 m. A box of mass 15 kg rests on the beam at its centre of mass.

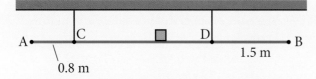

   (a) Copy the diagram, adding all the forces acting upon the beam.
   (b) Find the magnitude of the tension in both strings.
   (c) How far towards point B should the box be moved so that the tension in the string at D becomes twice the tension in the string at C?

4. A uniform beam ABCDEF of mass 4M kg is suspended horizontally in equilibrium by two light, inextensible strings at points A and E, as shown in the diagram. The distances between pairs of points are as follows: AB = x; BC = x; CD = 2x; DE = x; EF = 3x metres.

Exercise 4E...

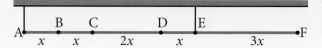

When a parcel of mass M kg is placed at point B on the beam, the tension in the string attached to point E is 17g N.
   (a) Copy the diagram, adding all the external forces acting upon the beam.
   (b) By taking moments about an appropriate point on the beam, form an equation and solve it to find M.
   (c) Find the tension in the string attached to point A.

5. A beam AB has mass 2 kg and length 6 m. It is held in equilibrium in a horizontal position by two vertical ropes attached to the beam. One rope is attached to A, the other to the point C on the beam where BC = 1.5 m, as shown in the diagram. The beam is modelled as a uniform rod, and the ropes as light strings.

   (a) Find:
       (i) the tension in the rope at C,
       (ii) the tension in the rope at A.
   A small load of mass 15 kg is attached to the beam at a point which is y metres from A. The load is modelled as a particle.
   (b) Given that the beam remains in equilibrium in a horizontal position, find, in terms of g and y, an expression for the tension in the rope at C.
   The rope at C will break if its tension exceeds 90 N. The rope at A cannot break.
   (c) Find the range of possible positions on the beam where the load could be attached without the rope at C breaking.

4.6   Non-Uniform Rods

A non-uniform rod does not have its mass distributed evenly along its length. Therefore, the centre of mass is not at the midpoint of the rod. It is not possible to model the rod's weight force as acting at the midpoint.

Instead, the weight force is modelled as acting at some other point on the rod. You may be asked to find the position of this centre of mass. Alternatively, you may be given the centre of mass and asked to find another unknown, such as a mass or the magnitude of a force.

....

Worked Example

8. A non-uniform beam AC of length 5 m and mass 25 kg has its centre of mass 3 m from point A. The beam is supported horizontally in equilibrium by two supports at points B and C, where B is 1 m from A.
   (a) Find the reaction forces at both supports.
   (b) State one modelling assumption that has been made.

   It is a good idea to draw a force diagram.

   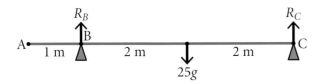

   (a) Take moments about point B:
   $$25g \times 2 = R_C \times 4$$
   $$50g = 4R_C$$
   $$\therefore R_C = 12.5g \text{ N or 123 N (3 s.f.)}$$

   Resolving vertically:
   $$R_B + R_C = 25g$$
   $$R_B + 12.5g = 25g$$
   $$\therefore R_B = 12.5g \text{ N or 123 N (3 s.f.)}$$

   R_B and R_C have the same magnitude. This makes sense, since the plank's weight force acts at the midpoint of B and C, so the weight is shared equally between the two supports.

   (b) The beam is modelled as a rod. The supports are assumed to be sharp.

....

In the next example we are asked to find the position of the rod's centre of mass.

....

Worked Example

9. A non-uniform plank AB has mass 20 kg and length 3 m. It is placed on two supports at A and B. A bucket of paint of mass 15 kg stands at a point P on the plank, where P is 0.6 m from B, as shown in the diagram.

The plank is horizontal and in equilibrium. The reaction at B is 1.5 times the reaction at A.
(a) Copy the diagram, showing the external forces acting on the plank.
(b) Find the magnitude of the reaction forces at A and B, giving answers in terms of g.
(c) Find the distance of the centre of mass of the plank from A.
(d) State a modelling assumption that has been made about each of:
   (i) the plank,
   (ii) the bucket of paint,
   (iii) the supports.

___

(a) The position of the plank's centre of mass is not known, so the weight force is marked with a distance x metres from point A.

(b) Resolving forces vertically:
$$R + 1.5R = 20g + 15g$$
$$2.5R = 35g$$
$$R = \frac{35g}{2.5} = 14g$$

Reaction at A: $14g$ N; Reaction at B: $21g$ N

(c) Take moments about A.
Note that distance AP = $3 - 0.6 = 2.4$ m
$$(20g \times x) + (15g \times 2.4) = (21g \times 3)$$
$$20gx + 36g = 63g$$
$$20gx = 27g$$
$$x = \frac{27}{20} = 1.35 \text{ m}$$

(d) (i) The plank is modelled as a rod.
   (ii) The bucket of paint is modelled as a particle.
   (iii) The supports are assumed to be sharp.

....

Exercise 4F

1. A non-uniform plank AB of length 1 metre rests on supports at A and B. If the magnitude of the reaction force at A is double the size of the reaction force at B, find the position of the plank's centre of mass, giving your answer in centimetres from A.

Exercise 4F...

2. A non-uniform beam of mass 12 kg rests horizontally in equilibrium on two supports at points P and Q. Its centre of mass lies 1 m from P and 3 m from Q. Find the magnitudes of the reaction forces at P and Q, giving your answers in terms of g.

3. The non-uniform beam ABCDEF shown in the diagram has a mass of 40 kg and a length of 5 m. It is supported at points B and E by ropes attached to a ceiling. The beam's centre of mass is at point C. A box of mass 2 kg rests on the beam at point D.

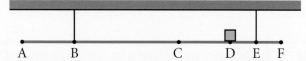

(a) Given that AB = CD = DF = 1 m, find the distance between points B and C.
(b) Copy the diagram, adding to it any forces acting upon the beam.
(c) Given that E is the midpoint of DF, find the tension in the rope attached to point E.
(d) Find the tension in the rope attached to point B.

4. A non-uniform plank AB has mass 90 kg and length 6 m. It rests on two supports at A and B. A crate of mass 72 kg stands at a point C on the plank, where C is 2 m from B, as shown in the diagram.

The plank is horizontal and in equilibrium. The reaction at B is twice the reaction at A.
(a) Copy the diagram, showing the external forces acting on the plank.
(b) Find the magnitude of the reaction forces at A and B.
(c) Find the distance of the centre of mass of the plank from A.
(d) State one modelling assumption that has been made.

5. A non-uniform rod AB of length L metres and mass M kg has its centre of mass at point G, which lies pL metres from point A (where $0 < p < 1$). The rod rests horizontally in equilibrium on supports at points A and B.

An additional mass of m kg rests on the rod at point P, which lies qL metres from A (where $p < q < 1$).

If the reaction forces at points A and B are R_A and R_B, show that:
$$R_A = \big((1 - p)M + (1 - q)m\big)g$$
and find R_B in terms of M, m, p, q and g.

4.7 Rods On The Point Of Tilting Or Turning

Consider a uniform rod of mass M kg resting on two supports, as in the diagram below. An additional mass m kg rests on the rod, towards one end.

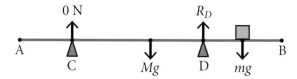

If the additional mass is heavy enough, it will cause the rod to tip or tilt, with point D as the pivot.

In this section we are interested in the case where the rod is **on the point of tilting**, i.e. it is still in equilibrium, but the reaction force at point C is zero, as the rod is on the point of lifting away from that support. This situation is known as **limiting equilibrium**.

Worked Example

10. The diagram shows a uniform plank of wood ABCDEF. The plank has a mass of 15 kg and a length of 5 m. The two supports shown are at points B and D, where AB = DF = 2 m. An additional mass m kg is placed at point E, where EF = 1 m.

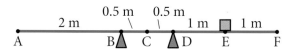

The plank is on the point of tilting about point D. Find m.

Draw a force diagram. Since the plank is on the point of tilting about point D, the reaction force from the support at B is zero.

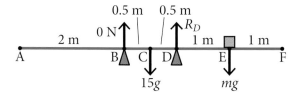

Take moments about D. In this way, the unknown reaction force R_D does not appear in the equation:

$$15g \times 0.5 = mg \times 1$$
$$mg = 7.5g$$
$$m = 7.5 \text{ kg}$$

In the following example, a beam is suspended by strings. The same principles are applied here. When the beam is on the point of turning about point C, the tension in the rope attached to point E becomes zero newtons, i.e. this rope becomes slack.

In this example, a distance is found.

Worked Example

11. The uniform beam shown in the diagram has a mass of 22 kg. It is suspended horizontally by two strings attached to points C and E. The beam's centre of mass is at point D, and the distance CD is 1 m. A box of mass 30 kg is placed at point B and the beam is on the point of tipping about point C.

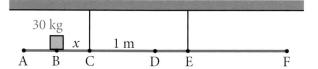

(a) Find the distance marked x between the box and point C.

(b) Find the magnitude of the tension in the string at C.

Draw a force diagram. Since the beam is on the point of tipping about C, the tension in the string at E is zero newtons.

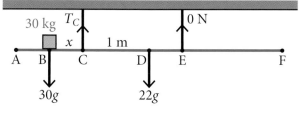

(a) Take moments about C, so that the unknown force T_C does not appear in the equation:

$$30g \times x = 22g \times 1$$
$$x = \frac{22}{30} = 0.733 \text{ m (3 s.f.)}$$

(b) Resolving forces vertically,
$$T_C = 30g + 22g$$
$$T_C = 52g = 510 \text{ N}$$

Exercise 4G

1. The horizontal beam ABCDEF shown in the diagram has its centre of mass at point D. A parcel of mass 4.5 kg rests at point B. The distance BC is 0.5 m. The beam is on the point of tilting about C.

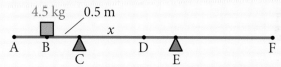

(a) State the magnitude of the reaction force on the beam at point E.

(b) Given that the beam has a mass of 1.5 kg, find the distance CD, marked x.

(c) The 4.5 kg parcel is replaced with a lighter parcel. Justifying your answer, state which one of the following statements is now correct:

(i) The beam is in equilibrium and the magnitudes of the two reaction forces at C and E are both greater than zero.

(ii) The beam is in equilibrium and the magnitude of the reaction force at E is zero.

(iii) The beam is no longer in equilibrium and tilts, with point C as the pivot.

2. A light beam ABCD of length x metres is supported in a horizontal position by two supports at points B and C, which are $\frac{x}{4}$ metres from either end, as shown in the diagram.

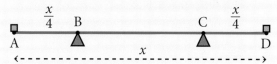

An object is placed on each end of the beam and the beam is on the point of tilting about C. Find the ratio of the masses of the objects at points A and D.

Exercise 4G...

3. A non-uniform straight rod PQ has length $4d$ metres and mass $6m$ kg. It is in equilibrium suspended horizontally by two strings at the points X and Y where PX = YQ = d and XY = $2d$. A particle of mass $2m$ is attached to the rod at Q. Given that the rod is on the point of tipping about Y, find the distance of the centre of mass of the rod from P, giving your answer in terms of d.

4. A uniform horizontal beam AB of mass 80 kg and length 10 m rests in equilibrium on supports at C and D, as shown in the diagram. AC = 2 m and DB = 3 m.

   (a) Draw a diagram showing all the external forces acting on the beam.
   (b) Find the magnitude of the reaction forces at C and D.

   An additional object of mass M kg is placed at B. The beam is now about to tilt.
   (c) Draw a second diagram showing all the external forces acting on the beam now.
   (d) Find M.
   (e) State two modelling assumptions that have been made in these calculations.

5. A uniform rod AB of mass M kg rests horizontally on two pivots C and D, such that AC = d_1 m, CD = d_2 m and DB = d_3 m as shown in the diagram. The centre of the rod is point E and the distance ED is marked x.

   A ⚫——— C ▲ ——— E x D ▲ ——— B ⚫
     d_1 d_2 d_3
           <------------>

   (a) Show that $x = \frac{1}{2}(d_1 + d_2 - d_3)$
   (b) Find, in terms of M, g, d_1, d_2 and d_3 the reactions at C and D.
   (c) An additional mass of m kg is placed at B so that the rod is on the point of tilting about pivot D.
      (i) State the size of the reaction force at pivot C.
      (ii) Find m in terms of M, d_1, d_2 and d_3.

4.8 Mixed Questions

The questions in the following exercise each combine two or more of the concepts and methods you have learnt in this chapter. These questions are of a standard to be expected in an examination paper.

Exercise 4H

1. A light rod ABC has length 2 m. A particle of mass 3 kg is attached to the rod at A and a particle of mass 2 kg is attached at B. The system is suspended from a ceiling by a light inextensible string attached at a point B on the rod, where AB is x metres. The system hangs in equilibrium with the rod horizontal as shown in the diagram. Find x.

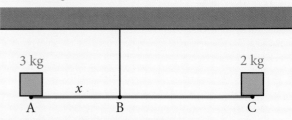

2. A non-uniform plank AB has length $5d$ metres and mass $8m$ kg. It is in equilibrium in a horizontal position resting on supports at the points P and Q where AP = $2d$ metres and AQ = $4d$ metres. A parcel of mass $6m$ is placed on the plank at B. The plank is on the point of tilting about Q.
   (a) By modelling the plank as a rod and the parcel as a particle, calculate, in terms of d, the distance of the centre of mass of the plank from A.
   (b) Explain briefly the significance of modelling the parcel as a particle.

3. On a construction site, a girder is being held in mid-air by a crane, on the end of a single inextensible chain, as shown in the diagram. The girder is non-uniform, has a mass of 200 kg and a length of 8 m.

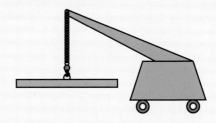

Exercise 4H...

The first attempt to lift the girder fails, since the chain hook is attached to the centre of the girder. In this case the girder experiences a clockwise moment of 100 N m. How far and in which direction should the hook be moved so that the girder remains horizontal as it is held?

4. A non-uniform beam AB, of mass M kg and length l metres, rests horizontally on two supports at points X and Y. The support at X is d_1 metres from A and the support at Y is d_2 metres from B. The beam's centre of gravity lies x metres from A, as shown in the diagram.

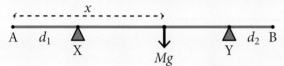

When an additional mass is placed at A, the beam is on the point of tilting about X. When the same mass is placed at B, the beam is on the point of tilting about Y.

(a) Show that $x = \dfrac{d_1 l}{d_1 + d_2}$

(b) Hence show that, if $d_1 = d_2$, then AB must be a uniform beam.

5. The diagram shows four family members, Alice, Ben, Caoimhe and Declan, sitting on a see-saw. Their positions are marked as A, B, C and D respectively. The see-saw has a mass of 80 kg, is non-uniform, horizontal and in equilibrium.

Alice, Ben and Caoimhe have masses of 30 kg, 16 kg and 40 kg respectively. The centre of mass is shown at point M and the distances between the points along the see-saw's length are shown on the diagram.

(a) Find Declan's mass in kilograms.

(b) How might the four people be related to each other? Explain your answer.

4.9 Summary

In this chapter you have learnt how to use moments for beams in simple static contexts, such as horizontal beams resting on pivots or supported by strings.

For a beam in equilibrium, clockwise moments are equal to anticlockwise moments.

You may need to take moments in combination with resolving forces in the vertical.

You may have to do this with uniform or non-uniform rods.

Using these techniques you can find the size of an unknown force, or the position of an object on a horizontal beam.

You will also come across beams that are in equilibrium but on the point of tilting about a pivot, or turning about one point if suspended by strings.

When a rod is on the point of tilting or turning about a point, the reaction force at any other support, or the tension in any other wire or string, becomes zero newtons.

Chapter 5
Rigid Bodies

5.1 Introduction

In the previous two chapters you learnt about moments acting upon **rods**, which are idealised one-dimensional **rigid bodies**.

When a force is applied to a rigid body, the force may cause the body to rotate, as well as experience a translational motion.

The turning effect of a force about a point depends on both the magnitude of the force and the position at which the force is applied.

The **moment** of a force is a measure of the force's ability to rotate the object on which it acts. It is the product of the magnitude of the force and the perpendicular distance from the point to the line of action of the force.

This chapter extends this work by considering two special types of rod: **ladders** and **hinged rods**.

Key words

- **Rigid body**: A one-, two- or three-dimensional object that does not bend or break.
- **Rod**: A one-dimensional rigid body. It has a length, but no width or height. The mass of a rod is concentrated along a line. In this chapter the types of rod considered are ladders and hinged beams.
- **Ladder**: A ladder can be modelled as a rod.
- **Hinged rod**: A hinged rod is a rod that is fixed at one end. This end of the rod cannot move, but the rod can turn about this point.
- **Equilibrium**: A rigid body is in equilibrium if it is not moving and not turning.
- **Moment**: The turning effect of a force about a point.

Before you start

You should know:

- The concept of a moment as the turning effect of a force.
- How to calculate the moment from an oblique force.
- How to calculate the resultant moment from a system of forces acting upon a rod.

- How to calculate the size of a force or the position of a force acting upon a rod, given that the rod is in equilibrium.
- How to calculate the size of a force or the position of a force acting upon a rod, given that the rod is on the point of turning.
- About the coefficient of friction and how to calculate the size of a friction force.
- About limiting equilibrium and that the maximum friction force is given by $F = \mu R$

Exercise 5A (Revision)

1. Consider the three forces acting upon the light rod shown in the diagram. Calculate the resultant moment about point B.

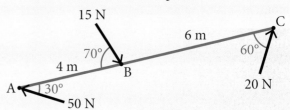

2. The non-uniform beam AB shown in the diagram has a length of 3.5 m and a mass of 8 kg. Its centre of mass is at the point M shown. The beam rests on two supports at points C and D. AC = 0.5 m and CM = MD = 1 m. A box of mass 20 kg is placed on the beam at a point that lies between points D and B.

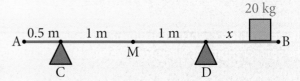

(a) Copy the diagram and add any forces acting upon the beam.

(b) Given that the beam is on the point of tilting about point D:

   (i) state the size of the reaction forces at points C and D;

   (ii) calculate the distance x between the box and point D.

Exercise 5A...

(c) State any modelling assumptions you have made in your calculations.

3. A crate of mass 30 kg is pulled along rough horizontal ground with a force of F newtons. Given that it accelerates uniformly from 0.1 m s^{-1} to 0.2 m s^{-1} in 5 seconds, find the coefficient of friction between the crate and the ground, giving your answer in terms of F and g in the form $\dfrac{aF - b}{cg}$, where a, b and c are integer constants to be found.

What you wIll learn

In this chapter you will learn how to:

- Model ladders as rods in simple static contexts and consider the forces and moments acting upon them.
- Use moments to solve problems involving ladders in limiting equilibrium.
- Model hinged rods and consider the forces and moments acting upon them.

In the real world...

Robots are becoming an increasingly common sight on factory production lines, replacing people for several good reasons: they are fast and accurate, they can work day and night without rest and they don't demand pay rises!

You may have seen videos of robotic arms rapidly assembling cars, or applying spray paint to them. In the manufacture and testing of pharmaceuticals robots can accurately combine the correct proportions of ingredients for a drug, shake or stir the vials in exactly the right way, and do much more.

In manufacturing, it is important for a robot to apply the correct force when attaching a component, or the correct moment when tightening a bolt, to avoid damage. The correct forces and moments must also be applied to the robotic arm itself to swing it into the correct position at speed. All of this requires careful computer programming and a degree of artificial intelligence.

Occasionally mistakes occur. In July 2022, a chess-playing robot malfunctioned when playing against a 7-year-old boy called Christopher in a tournament in Moscow. The robot appeared to grab Christopher's

finger, breaking it. Reports suggest that Christopher took his move more quickly than the robot was programmed to expect, confusing the machine!

5.2 Uniform Ladders

In questions involving ladders, it is common to take moments about the foot of the ladder. In this way, any forces passing through that point can be eliminated from the moments equation.

It is important to understand how the moments from the other forces are calculated. The following example demonstrates this.

....

Worked Example

1. The uniform ladder AC shown in the diagram has a mass of M kg and a length of $2l$ metres. Considering moments about point A, find the size of the moments provided by:
   (a) the force of R newtons shown, giving your answer in term of R, l and θ.
   (b) the weight force of Mg newtons shown, giving your answer in term of M, g, l and θ.

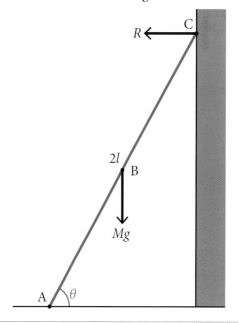

Recall that the magnitude of the moment from an oblique force acting upon a rod is $Fd \sin \theta$, where:

- F is the size of the force;
- d is the distance from the point at which the force acts to the point about which moments are being taken; and
- θ is the angle between the force and the rod.

The diagram has been redrawn to show appropriate lengths and angles.

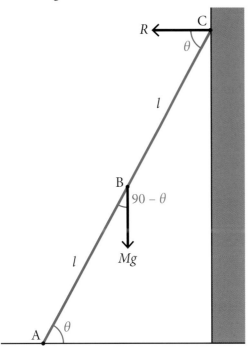

(a) The force of R newtons acts at an angle $\theta°$ to the ladder. The point at which it acts on the ladder is $2l$ metres from A. Therefore, the moment from this force is:

$$R \times 2l \times \sin \theta$$
$$= 2Rl \sin \theta \text{ anticlockwise}$$

(b) The force of Mg newtons acts at an angle $(90 - \theta)°$ to the ladder. The point at which it acts is l metres from A. Therefore, the moment from this force is:

$$Mg \times l \times \sin(90 - \theta)$$
$$= Mgl \cos \theta \text{ clockwise,}$$
$$\text{since } \cos \theta \equiv \sin(90 - \theta)$$

Note: There are alternative methods for finding these two moments. For example each force could be resolved into two components, parallel and perpendicular to the ladder.

Generally, a ladder standing on horizontal ground and leaning against a vertical wall experiences two reaction forces: one from the ground, acting vertically upwards on the foot of the ladder; the second from the wall, acting horizontally at the top of the ladder.

If the ground is rough, there is a frictional force acting on the foot of the ladder, towards the wall. Additionally, if the wall is rough, there is a second frictional force acting

at the top of the ladder, upwards.

If it is unclear in which direction these frictional forces should act, consider the motion of the ladder if it slipped. Both frictional forces oppose the motion.

The following example shows a uniform ladder on rough ground, leaning against a rough wall.

....

Worked Example

2. The uniform ladder shown in the diagram stands on rough ground and leans against a rough wall. It experiences a normal reaction force of $30g$ N from the ground, acting at its foot.

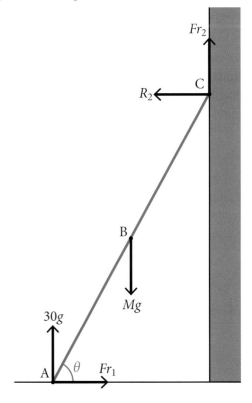

The frictional force at the foot of the ladder is one fifth of the size of the normal reaction. The frictional force at the top of the ladder is one sixth of the size of the normal reaction from the wall. Calculate:

(a) the ladder's mass
(b) the angle θ between the ladder and the horizontal ground.

The force diagram is already given. In an exam setting it would be permissible to write the values of the forces onto the diagram given as you calculate them.

(a) Since the frictional force is ⅕ of the normal reaction at the foot of the ladder:

$$Fr_1 = 6g$$

Next, resolve forces in the horizontal:

$$Fr_1 = R_2$$
$$\therefore R_2 = 6g$$

Since the frictional force at the top of the ladder is ⅙ of the normal reaction:
$$Fr_2 = g$$

Resolving forces in the vertical gives:
$$30g + g = Mg$$
$$Mg = 31g$$
$$M = 31 \, \text{kg}$$

(b) Let the ladder's length be l metres.
Taking moments about point A (the foot of the ladder):

$$Mg \times \frac{l}{2} \times \cos\theta$$
$$= (R_2 \times l \times \sin\theta) + (Fr_2 \times l \times \cos\theta)$$

Use values for M, R_2 and Fr_2:
$$\frac{31gl}{2} \cos\theta = (6gl \sin\theta) + (gl \cos\theta)$$

Divide by gl and by $\cos\theta$:
$$\frac{31}{2} = 6\tan\theta + 1$$
$$31 = 12\tan\theta + 2$$
$$\tan\theta = \frac{29}{12}$$
$$\theta = \tan^{-1}\left(\frac{29}{12}\right) = 67.5° \, (3 \text{ s.f.})$$

> **Note:** Frictional forces are considered in this example, but it is not possible to use the formula $Fr = \mu R$, since the ladder is not moving or in **limiting equilibrium**. Limiting equilibrium is considered in section 5.4.

Example 3 below involves an additional object – in this case a painter – on a ladder. The question also features a rope attached to the foot of the ladder, providing an additional force.

Worked Example

3. A uniform ladder of mass 20 kg rests in equilibrium with its base on a smooth horizontal floor and its top against a smooth vertical wall. The base of the ladder is 1.5 m from the wall and the top of the ladder is 2 m from the floor. A painter of mass 80 kg climbs three-fifths of the way up the ladder. The ladder is kept in equilibrium by a rope attached to the base of the ladder and to a hook on the wall 1.125 m above the floor.

(a) Draw a diagram, showing all the forces acting upon the ladder.

(b) Find the tension T in the rope, giving an exact answer.

(c) What modelling assumptions have been made in answering this question?

(a) The rope is marked on the diagram and the tension in the rope is T N. Since the ground and wall are both smooth, there are no friction forces. The centre of the ladder is C and the painter is at P. The hook on the wall is marked as H. Distances between points on the ladder have been calculated.

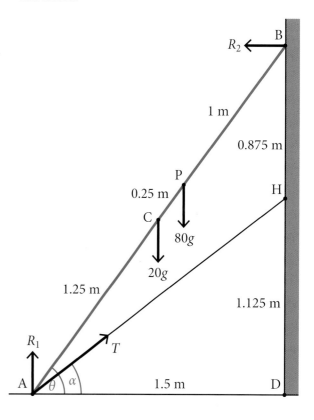

(b) Using triangle ABD we can calculate:
$$\sin\theta = \frac{4}{5}, \cos\theta = \frac{3}{5}$$

By Pythagoras' Theorem in triangle AHD, the length of the rope is 1.875 m and we can calculate:
$$\sin\alpha = \frac{3}{5}, \cos\alpha = \frac{4}{5}$$

Taking moments about A:

$R_2 \times 2.5 \times \sin\theta$
$\qquad = (20g \times 1.25 \times \cos\theta) + (80g \times 1.5 \times \cos\theta)$
$2R_2 = 15g + 72g$
$2R_2 = 87g$
$\quad R_2 = 43.5g$

Resolving in the horizontal:

$T\cos\alpha = R_2$
$\quad \dfrac{4}{5}T = R_2$
$\qquad T = \dfrac{435}{8}g$

(c) The rope is modelled as a light, inextensible string. The painter is modelled as a particle. The ladder is modelled as a rod.

....

If a rod leans against a post or against a corner, the normal reaction force at the point of contact acts perpendicular to the rod, as shown in the following example.

....

Worked Example

4. A 6-metre uniform ladder stands on rough ground and leans against a 4-metre high barn, so that the top one metre of the ladder protrudes above the barn, as shown in the diagram.

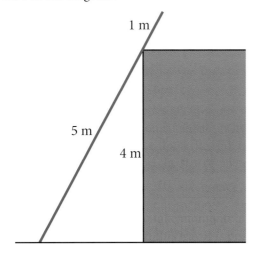

The mass of the ladder is 25 kg.
(a) Draw a diagram showing all the external forces acting upon the ladder.
(b) Find the size of the normal reaction force that the barn exerts on the ladder, giving your answer in terms of g.

(a)

R₂ diagram showing forces: R_2 at C, 1 m to D, 2 m, 4 m, 3 m, $25g$ at B, R_1 and Fr_1 at A, angle θ, 3 m at G.

(b) Using right-angled triangle AGC, Distance AG has been calculated as 3 m.

Take moments about the foot of the ladder, point A. Note that R_2 acts perpendicular to the ladder, so the moment from this force is easily calculated as the magnitude of the force multiplied by the perpendicular distance of 5 m.

$$25g \times 3 \times \cos\theta = R_2 \times 5$$

From the diagram it can be seen that $\cos\theta = \dfrac{3}{5}$

$$\therefore 75g \times \frac{3}{5} = 5R_2$$
$$5R_2 = 45g$$
$$R_2 = 9g \text{ N}$$

....

Exercise 5B

1. Considering the uniform ladder shown in the diagram, calculate the size and the sense of the moments about point A, due to the two forces shown.

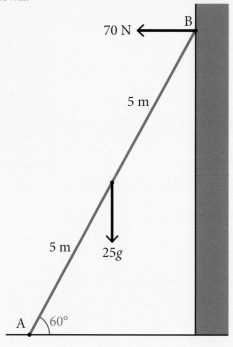

2. A small uniform ladder leans in equilibrium against a vertical cupboard at an angle of 70° to the horizontal, as shown in the diagram.

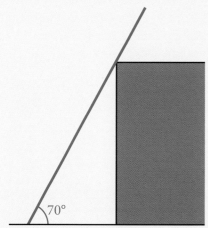

Given that the floor and cupboard are smooth, that the ladder has a mass of 7 kg and that one quarter of the ladder's length extends above the cupboard, find the magnitude of the reaction force that the cupboard exerts upon the ladder.

Exercise 5B...

3. A uniform ladder AB of weight 150 N and length 8 m rests in equilibrium. Its end A rests on rough horizontal ground and B rests against a smooth vertical wall. AB makes an angle $\tan^{-1}\left(\dfrac{12}{5}\right)$ with the horizontal as shown below.

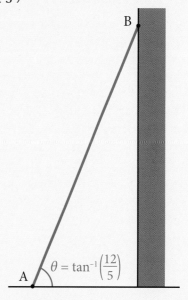

   (a) Draw a diagram showing all the external forces acting on the ladder.
   (b) By resolving forces in the vertical, find the magnitude of the normal reaction at point A.
   (c) By taking moments about point A, find the magnitude of the normal reaction at B.
   (d) Find the size of the frictional force acting at the foot of the ladder.

4. A uniform ladder stands on rough ground and leans against a rough wall. It experiences a normal reaction force of $24g$ N from the ground, acting at its foot.
   (a) Draw a force diagram, showing all the external forces acting upon the ladder.

   The frictional force at the foot of the ladder is one third of the size of the normal reaction. The frictional force at the top of the ladder is one quarter of the size of the normal reaction from the wall.
   (b) Calculate, giving answers in terms of g:
     (i) the size of the frictional force acting at the foot of the ladder;
     (ii) the size of the normal reaction force from the wall, acting at the top of the ladder;

Exercise 5B...

    (iii) the size of the frictional force acting at the top of the ladder.

  (c) Calculate

    (i) the ladder's mass;

    (ii) the angle θ between the ladder and the horizontal ground.

5. A uniform ladder of mass 8 kg rests in equilibrium with its base on a smooth horizontal floor and its top against a smooth vertical wall. The base of the ladder is 1 m from the wall and the top of the ladder is 2 m from the floor. The ladder is kept in equilibrium by a light inextensible string attached to the base of the ladder and to a point on the wall exactly halfway between the top of the ladder and the floor.

  (a) Find the tension in the string, giving an exact answer.

  (b) What modelling assumptions have been made in answering this question?

6. A uniform ladder of mass 10 kg and length 4 m rests with one end on a smooth horizontal floor and the other end against a smooth vertical wall. The ladder is kept in equilibrium at an angle of θ to the horizontal where $\tan\theta = 2$, by a light inextensible string attached to the base of the ladder and to the base of the wall, at a point vertically below the top of the ladder. A man of mass 100 kg ascends the ladder.

  (a) If the string will break when the tension exceeds 400 N, find how far up the ladder the man can ascend before this occurs.

  (b) What tension must the string be capable of withstanding if the man is to reach the top of the ladder?

7. Paul the window cleaner places his uniform ladder on rough horizontal ground and against the rough, vertical wall of a house. The ladder has a length of 4.5 m and a mass of 20 kg. It is placed at an angle θ to the ground, where $\tan\theta = \dfrac{8}{3}$. The ladder experiences reaction forces of magnitude R_1 and R_2 newtons at the foot and top of the ladder respectively. It also experiences friction forces of magnitude Fr_1 and Fr_2 newtons at the foot and top of the ladder respectively. Paul has a mass of 60 kg and he climbs one-third of the way up the ladder.

Exercise 5B...

  (a) Draw a diagram showing all the forces acting on the ladder.

  (b) Resolve forces in the vertical to obtain Fr_2 in terms of R_1 and g.

  (c) Take moments about the foot of the ladder to show that $150g = 3R_1 - 8R_2$

  (d) Given that the ratio of the reaction forces at the foot and top of the ladder is 4:1, find the magnitude of these reaction forces, giving answers in terms of g.

  (e) State any modelling assumptions that have been made.

5.3 Non-Uniform Ladders

A **non-uniform** ladder is a ladder whose mass is not evenly distributed along its length. It can be modelled as a rod whose centre of mass is not halfway along its length. In a question involving a non-uniform ladder, you may be asked to calculate the position of the centre of mass or use the position of the centre of mass given to find another quantity.

....

Worked Example

5. The diagram shows a non-uniform ladder ABC of mass 9 kg and length 4 m. Point B is the ladder's centre of mass. It stands on rough ground and leans against the smooth wall of a house. The top of the ladder is 3 m above ground level. The frictional force at the foot of the ladder is 35 N.

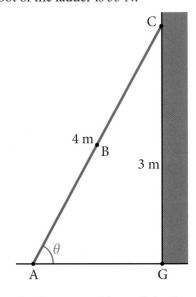

  (a) Copy the diagram, adding all the forces acting upon the ladder.

(b) Find AB, the distance of the ladder's centre of mass from the foot of the ladder.

(a)

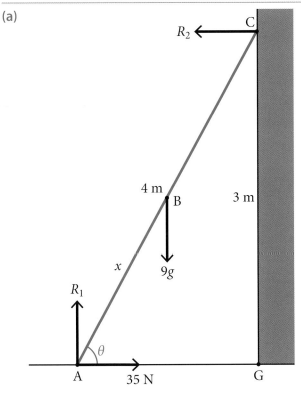

(b) Resolving forces in the horizontal gives
$R_2 = 35\,\text{N}$

Taking moments about point A:
$$9gx \cos \theta = 35 \times 3 = 105$$
$$x = \frac{105}{9g \cos \theta}$$

Considering the right-angled triangle ACG:
$$AG = \sqrt{4^2 - 3^2} = \sqrt{7}$$
$$\cos \theta = \frac{\sqrt{7}}{4}$$
$$\therefore x = \frac{105}{9g \times \sqrt{7}/4} = 1.80\,\text{m (3 s.f.)}$$

Given the position of the centre of mass of a non-uniform ladder, you may be asked to work out another value, such as the angle of the ladder, or its mass. In the next example the angle must be calculated.

Worked Example

6. A non-uniform ladder AB has mass 12 kg and length 6 m. Its centre of mass is 2.8 m from end A. The ladder stands with end A on rough horizontal ground and it leans against a smooth vertical wall. The wall exerts a normal reaction force of $4g$ N at the top of

the ladder. Find the size of the angle the ladder makes with the ground.

Although the question does not ask for a force diagram, it is a good idea to draw one.

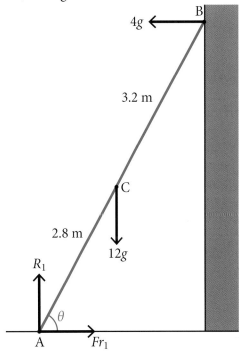

The centre of mass has been labelled C. Since we are told AC = 2.8 m, BC can be calculated as 3.2 m. The angle the ladder makes with the horizontal is labelled θ.

Taking moments about point A:
$$12g \times 2.8 \times \cos \theta = 4g \times 6 \times \sin \theta$$
$$33.6g \cos \theta = 24g \sin \theta$$
$$\tan \theta = \frac{33.6g}{24g} = \frac{7}{5}$$
$$\theta = \tan^{-1}\left(\frac{7}{5}\right) = 54.5° \text{ (3 s.f.)}$$

Exercise 5C

1. A non-uniform ladder has mass 20 kg and length 6 metres. Its centre of mass is 3.2 m from the foot of the ladder. The ladder is in equilibrium with end A on rough horizontal ground. It leans against a smooth vertical wall at an angle of 72° to the horizontal. Find:
   (a) the magnitude of the normal reaction force at the foot of the ladder;
   (b) the magnitude of the normal reaction force at the top of the ladder.

Exercise 5C...

2. The diagram shows a non-uniform ladder ABC of length 5 m. Point B is the ladder's centre of mass. The foot of the ladder is 2 m horizontally from the vertical wall.

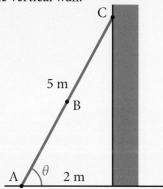

The ladder stands on rough ground and leans against a smooth wall. Its mass is 20 kg and the frictional force at the foot of the ladder is 40 N.
   (a) Copy the diagram, adding all the forces acting upon the ladder.
   (b) Find AB, the distance of the ladder's centre of mass from the foot of the ladder.

3. A non-uniform ladder has mass M kg and length l metres. The ladder's centre of mass is positioned $0.9l$ metres from its foot. The ladder stands with its foot on rough horizontal ground and it leans against a smooth vertical wall at an angle of $\theta°$ to the horizontal. The wall exerts a normal reaction force of $6g$ N at the top of the ladder.
   (a) Show that $M = \dfrac{20}{3}\tan\theta$
   (b) Hence find the value of M, given that the gradient of the ladder is ⁹⁄₅.

4. A non-uniform ladder AB has mass 10 kg and length 5.5 m. Its centre of mass is 2.2 m from end A. The ladder stands with end A on rough horizontal ground and it leans against a smooth vertical wall.

   The wall exerts a normal reaction force of $\dfrac{7g}{2}$ N at the top of the ladder.

   Find the size of the angle the ladder makes with the ground, giving your answer in the form $\tan^{-1}\left(\dfrac{a}{b}\right)$, where $\dfrac{a}{b}$ is a fraction to be found in its simplest form.

Exercise 5C...

5. A man of mass M kg climbs a non-uniform ladder AB of mass m kg. The ladder is in equilibrium at an angle of $\theta°$ to the ground. The ladder's lower end A is on rough horizontal ground and its upper end B rests against a smooth vertical wall. The ladder's centre of mass is ²⁄₅ of the distance from A to B. The man stands at a point that is ³⁄₅ of the distance from A to B. At this point, the frictional force at the foot of the ladder is ⅕ of the normal reaction force.
   (a) Draw a diagram showing all the forces acting on the ladder.
   (b) Show that $\tan\theta = \dfrac{2m + 3M}{m + M}$
   (c) Hence find the angle the ladder makes with the horizontal if the man's mass is double the ladder's mass.
   (d) What modelling assumptions have been made in answering this question?

6. A non-uniform ladder AB of mass M kg rests on a post, as shown in the diagram. The foot of the ladder is on rough horizontal ground and the ladder is inclined to the horizontal at an angle of 60°. The length of the ladder is L metres. Its centre of mass is at point C, which is $\dfrac{L}{3}$ metres from end A. The point at which the ladder makes contact with the post is $\dfrac{L}{3}$ metres from end B.

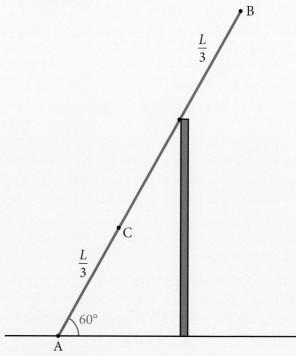

Exercise 5C...

(a) Copy the diagram, adding all external forces acting upon the ladder.

(b) Show that the magnitude of the reaction force at the post is $\dfrac{Mg}{a}$, where a is a constant integer to be found.

(c) Find the magnitude of the frictional force acting upon the ladder at point A, giving an exact answer in terms of M and g.

5.4 Ladders in Limiting Equilibrium

In the work on friction in AS Applied Mathematics you learnt about limiting equilibrium. You learnt that, for an object that is moving or on the point of moving on a plane, the magnitude of the frictional force is given by μR, where μ is the coefficient of friction and R is the size of the normal reaction force.

This model for the magnitude of the frictional force can be applied to ladders.

....

Worked Example

7. A uniform ladder AC of length $2l$ m rests in limiting equilibrium with its top end against a rough vertical wall and its lower end on a rough horizontal floor, as shown. If the coefficient of friction at the foot and top of the ladder are ¼ and ⅔ respectively, find the angle θ that the ladder makes with the floor.

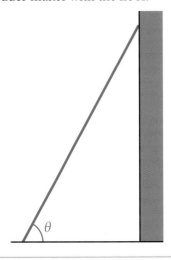

Let the midpoint of the ladder be B and its mass be M kg.

Let the magnitudes of the reaction forces from the ground and the wall be R_1 and R_2 respectively.

Draw a force diagram for the ladder. Since the ladder is in limiting equilibrium, the magnitudes of the frictional forces at the foot and top of the ladder can be calculated in terms of the reaction forces.

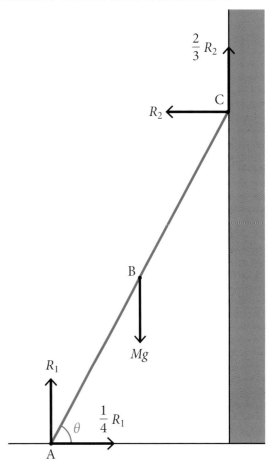

Resolving vertically,

$$R_1 + \frac{2}{3}R_2 = Mg \qquad (1)$$

Resolving horizontally,

$$R_2 = \frac{1}{4}R_1 \Rightarrow R_1 = 4R_2 \qquad (2)$$

Eliminating R_1:

$$4R_2 + \frac{2}{3}R_2 = Mg \Rightarrow \frac{14}{3}R_2 = Mg \qquad (3)$$

Taking moments about A:

$$Mg \times l \cos\theta = R_2 \times 2l \sin\theta + \frac{2}{3}R_2 \times 2l \cos\theta$$

$$Mgl \cos\theta = 2R_2 l \sin\theta + \frac{4}{3}R_2 l \cos\theta$$

$$Mg = 2R_2 \tan\theta + \frac{4}{3}R_2$$

Substituting for Mg from (3):

$$\frac{14}{3}R_2 = 2R_2 \tan\theta + \frac{4}{3}R_2$$

$$\frac{14}{3} = 2\tan\theta + \frac{4}{3}$$

$$\tan\theta = \frac{5}{3}$$

$$\theta = 59.0° \text{ (3 s.f.)}$$

The following example shows a uniform ladder on rough ground, leaning against a smooth wall, in limiting equilibrium.

Worked Example

8. A uniform ladder of length $2l$ metres and mass 20 kg rests in equilibrium with its upper end against a smooth vertical wall and its lower end on rough horizontal ground, with coefficient of friction μ. The normal reaction forces at the ground and wall are R and S newtons respectively and F is the frictional force at the ground. The ladder makes an angle of 45° with the ground. Find:
   (a) the magnitude of R, S and F; and
   (b) the smallest possible value of μ.

No diagram is provided here, so it is important to draw one. The endpoints of the ladder have been labelled A and C and the midpoint B.

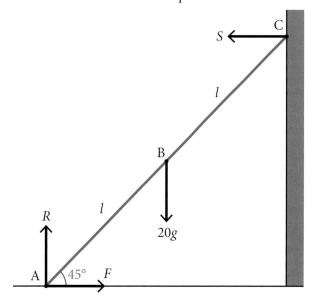

(a) Resolving in the vertical,
$$R = 20g \qquad (1)$$

Resolving in the horizontal,
$$S = F \qquad (2)$$

Taking moments about A:
$$20g \times l \times \cos 45 = S \times 2l \times \sin 45$$

Divide by $2l$ and by $\cos 45$:
$$10g = S\tan 45$$
$$S = 10g$$

(b) We know that the maximum size of the frictional force F is μR.

$$F \geq \mu R$$
$$\Rightarrow \mu \leq \frac{F}{R}$$
$$\mu \leq \frac{10g}{20g}$$
$$\mu \leq 0.5$$

Exercise 5D

1. A uniform ladder rests in limiting equilibrium with its top end against a rough vertical wall and its lower end on a rough horizontal floor. If the coefficient of friction at the foot and top of the ladder are ⅕ and ¾ respectively, find the angle that the ladder makes with the floor.

2. A uniform ladder AB of mass 30 kg and length 6 m rests in limiting equilibrium. Its end A rests on rough horizontal ground and B rests against a smooth vertical wall. AB makes an angle $\tan^{-1}\left(\frac{24}{7}\right)$ with the horizontal as shown below.

Exercise 5D...

The coefficient of friction between A and the ground is μ.

(a) Draw a diagram showing all the external forces acting on the ladder.

(b) Show that the normal reaction at B is $\dfrac{35}{8}g$ N.

(c) Find μ.

3. A uniform ladder rests in limiting equilibrium with its top end against a rough vertical wall. The coefficient of friction between the top of the ladder and the wall is μ. The foot of the ladder is on rough horizontal ground. The coefficient of friction between the foot of the ladder and the ground is λ. If the ladder makes an angle of $\theta°$ with the floor, show that $\tan\theta = \dfrac{1 - \mu\lambda}{2\lambda}$

4. A uniform ladder of length $2l$ metres and mass 30 kg rests in equilibrium with its upper end against a smooth vertical wall and its lower end on rough horizontal ground, with coefficient of friction μ. The normal reaction forces at the ground and wall are R_1 and R_2 newtons respectively and F is the frictional force at the ground. The ladder makes an angle of 60° with the ground. Find:

(a) the magnitude of R_1, R_2 and F; and

(b) the smallest possible value of μ.

5. A man of mass M kg climbs a non-uniform ladder AB of mass m kg and length L metres. The ladder's lower end A is on rough horizontal ground and its upper end B rests against a smooth vertical wall. The ladder is in limiting equilibrium at an angle of $\theta°$ to the ground. The ladder's centre of mass is pL metres from A, where $0 < p < 1$. The man stands at a point that is qL metres from A, where $0 < q < 1$. The coefficient of friction between the ladder and the ground is μ.

(a) Show that $\tan\theta = \dfrac{pm + qM}{\mu(m + M)}$

(b) The ladder's centre of mass is at a point 45% of the distance from A to B. The man stands ¾ of the way up the ladder. Given also that $\mu = 0.15$ and that the masses of the man and ladder are in the ratio 3:1, find $\tan\theta$ as a simplified fraction.

Exercise 5D...

6. Two uniform ladders L and R with lengths 10 m and 17 m rest in equilibrium against opposite sides of a smooth vertical wall, as shown in the diagram. The tops of the ladders rest against the wall at the same height. Both ladders are made of metal weighing 1 kg per metre of length. The foot of Ladder L is 6 m from the wall. The feet of both ladders rest on rough horizontal ground.

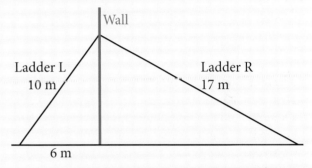

(a) Copy the diagram, showing all the forces acting on both ladders.

(b) Find how far the foot of Ladder R lies from the wall.

The coefficient of friction between ladder L and the ground is μ. The coefficient of friction between ladder R and the ground is λ.

(c) Given that both ladders are on the point of slipping, find μ and λ.

5.5 Hinged Rods

A rod that is hinged at a point on a wall experiences a reaction force from the wall that can be considered as having two components:

- Since the rod is in contact with the wall, it experiences a normal reaction force, which acts at right angles to the wall.

- Secondly, the hinge provides support for the rod, so there is a component of the reaction force in the vertical, as if the rod were resting on a pivot at that point.

If you are asked to find the size or magnitude of the reaction force, find both of these components and then use Pythagoras' Theorem to find the resultant. If you are asked for the direction at which the reaction force acts, trigonometry is required.

It is common to use X for the horizontal component of the reaction force and Y for the vertical component.

Modelling assumptions

In all the examples and questions in this section, all objects are modelled as rods.

Strings can be assumed to be light and inextensible, unless otherwise stated in the question.

The hinge is assumed to be smooth, which means that there is no friction in the hinge. If this is not stated in the question, you may state that the hinge is smooth if asked for modelling assumptions.

All forces acting on the rods are assumed to be coplanar, i.e. they act in the same plane.

....

Worked Example

9. A uniform beam AB of mass $5m$ kg and length 3 metres is smoothly hinged at end A to a fixed point on a vertical wall as shown in the diagram. The beam is maintained in a horizontal position by a light inelastic string attached to a point D on the wall and a point C on the beam. The string is inclined to the beam at an angle $\theta°$. AC = 2.24 m and AD = 1.68 m.

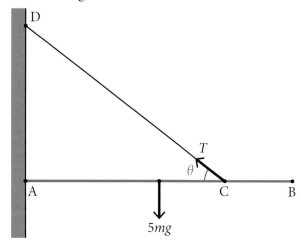

(a) Copy the diagram, adding any additional forces acting on the beam.

(b) Find the magnitude of the reaction acting on the beam at the hinge and give the direction at which it acts to the horizontal.

(a) The two components X and Y of the reaction force at the hinge have been added to the diagram.

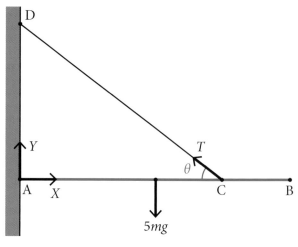

(b) Using right-angled triangle CAD:

$$CD = \sqrt{1.68^2 + 2.24^2} = 2.8$$

$$\sin \theta = \frac{1.68}{2.8} = \frac{3}{5}$$

$$\cos \theta = \frac{2.24}{2.8} = \frac{4}{5}$$

Take moments about point A. Neither X nor Y appear in the equation, since the lines of these forces pass through point A.

$$T \times 2.24 \sin \theta = 5mg \times 1.5$$

$$1.68T = 7.5mg$$

$$T = \frac{7.5mg}{1.68} = \frac{125}{28}mg$$

Resolving in the horizontal:

$$X = T \cos \theta = \frac{125}{28}mg \times \frac{4}{5} = \frac{25}{7}mg$$

Resolving in the vertical:

$$Y + T \sin \theta = 5mg$$

$$Y + \frac{125}{28}mg \times \frac{3}{5} = 5mg$$

$$Y + \frac{75}{28}mg = 5mg$$

$$Y = \frac{65}{28}mg$$

Find the magnitude of the reaction force using Pythagoras' Theorem. Add the two components together using vector addition, as in the diagram:

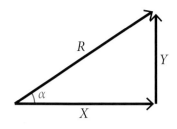

$$R = \sqrt{\left(\frac{65}{28}mg\right)^2 + \left(\frac{25}{7}mg\right)^2} = 41.7m \text{ (3 s.f.)}$$

The direction of the reaction force to the horizontal is shown as α in the diagram. This can be found using trigonometry:

$$\tan \alpha = \left(\frac{65}{28}mg\right) \div \left(\frac{25}{7}mg\right) = \frac{13}{20}$$

$$\alpha = \tan^{-1}\left(\frac{13}{20}\right) = 33.0°$$

Exercise 5E

1. A uniform rod PQ is hinged at point P on a vertical wall and is held in equilibrium at an angle of 40° to the horizontal by a cable. The cable is attached at right angles to the rod at point Q, as shown, and the tension in the cable has magnitude T newtons. The rod has a length of 4 m and a mass of 10 kg.

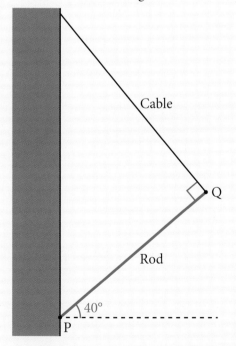

(a) Copy the diagram, adding all the external forces acting on the rod.
(b) By taking moments about point P, find the value of T.

Exercise 5E...

2. A uniform rod AB of mass 10 kg is hinged smoothly to a wall at end A. It is held in horizontal equilibrium by a force F newtons acting at B, which acts at 45° to the rod, as shown in the force diagram. The rod's weight is shown, as are the two components of the reaction force at the hinge.

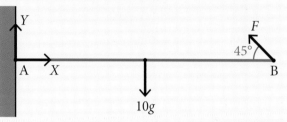

(a) By taking moments about point A, find the magnitude of F.
(b) Find the magnitude and direction of the reaction at the hinge.

3. The diagram shows a uniform drawbridge AB, of mass 900 kg, that can rise to allow boats to pass through it. The drawbridge rotates about its end A, which is hinged to a wall. A horizontal cable is attached at point C on the drawbridge, providing a horizontal force of magnitude 5000 N. The distances AC and CB are in the ratio 3:1.

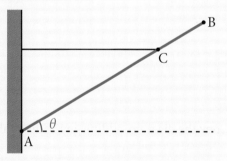

(a) Copy the diagram showing all the forces acting upon the drawbridge.
(b) The drawbridge is at an angle of $\theta°$ to the horizontal. Given that the bridge is rising, show that $\tan \theta > \frac{3g}{25}$
(c) State any modelling assumptions used in your calculations.

Exercise 5E...

4. Ben lives in a houseboat. To save space his bed folds up against the wall. It is hinged at point A. The diagram shows the bed at an angle $\alpha°$ to the horizontal. Ben is folding the bed up against the wall, pushing at point B.

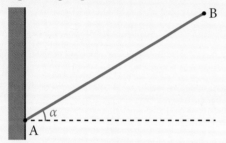

   (a) Which of the following is the best way for Ben to push at point B?
   (i) horizontally;
   (ii) vertically;
   (iii) perpendicular to the bed?
   Justify your answer by referring to an expression for the size of the moment that results when Ben pushes at point B.
   (b) State any modelling assumptions you have made.

5. Finn has a lot of recycling to put in his wheelie bin, so he uses a piece of string, attached to the bin lid at point C and to a hook on the wall, to keep the lid open. The diagram shows the bin lid being held open at an angle of $\alpha°$ to the horizontal. The string is at an angle of $\theta°$ to the bin lid. The lid has a mass of M kg and a length of l metres. Point C is a distance $\dfrac{3l}{4}$ metres from A.

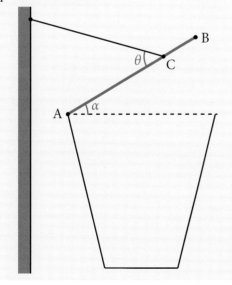

Exercise 5E...

   (a) Copy the diagram, adding any forces acting upon the bin lid.
   (b) When $\alpha = 30°$ and $\theta = 45°$, the tension in the string is 40 N. Show that the mass of the lid is 5.0 kg to 2 significant figures.
   (c) Find an expression for T in terms of M, g, α and θ.
   (d) The string will break if the tension is greater than 60 N. If θ remains fixed at 45°, find the largest tension possible in the string and state if the string could break.
   (e) State any modelling assumptions made in answering part (b).

6. A uniform beam AB of mass 20 kg and length 6 metres is smoothly hinged at end A to a fixed point on a vertical wall as shown in the diagram. The beam is held in a horizontal position by a light inelastic string attached to a point D on the wall and a point C on the beam. The string is inclined to the beam at an angle $\theta°$. AC = 3.5 m and AD = 2.5 m.

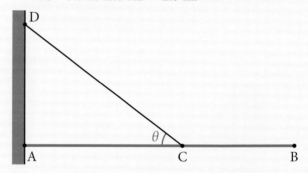

   (a) Copy the diagram, adding all the forces acting on the beam.
   (b) Find the magnitude of the reaction force acting on the beam at the hinge, giving your answer to the nearest newton.
   (c) Find the direction to the horizontal at which the reaction force acts upon the beam.

7. A uniform rod AB of mass M kg and length 1 metre is smoothly hinged at end A to a fixed point on a vertical wall as shown in the diagram. The beam is maintained in a horizontal position by a light inelastic string attached to a point D on the wall and a point C on the beam, where AC = 0.75 m. The string is inclined to the beam at an angle of $\theta°$.

Exercise 5E...

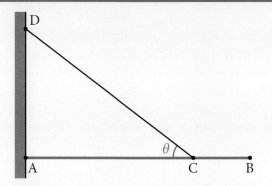

When $\theta = 45°$, the tension in the string is $\dfrac{\sqrt{2}}{2}T$ newtons. The horizontal and vertical components of the reaction force at the hinge are $2g$ and Y_1 newtons respectively.

(a) Show that $T = 4g$ newtons and find M in kilograms.

(b) Find Y_1 in newtons, giving your answer in terms of g.

When $\theta = 30°$, the tension in the string is T newtons. The horizontal and vertical components of the reaction force at A are X and Y newtons respectively.

(c) Find X and Y in newtons, giving your answers in terms of g.

5.6 Summary

In questions involving ladders, it is common to take moments about the foot of the ladder. This ensures that any forces passing through that point (typically a normal reaction force and possibly a frictional force) do not appear in the moments equation. For any other forces acting upon the ladder, the moment can be calculated using $Fd \sin \theta$, where:

- F is the magnitude of the force;
- d is the distance between the point at which the force acts and the point about which moments are taken;
- θ is the angle between the line of the force and the ladder.

Questions may involve non-uniform ladders. In these questions, you may be asked to calculate the position of the centre of mass, or use the position of the centre of mass given to find another unknown.

For a ladder that is on the point of slipping, the frictional force at the foot of the ladder can be calculated using $F_r = \mu R$, where μ is the coefficient of friction between the ladder and the ground and R is the normal reaction force at that point.

The ladder may be leaning against a rough wall, in which case the frictional force at the top of the ladder is calculated in a similar way.

If a rod is hinged to a wall, the rod typically experiences a reaction force comprising vertical and horizontal components.

For questions involving hinged rods, it is common to take moments about the point at which the rod is hinged. This ensures that the horizontal and vertical components of the reaction force do not appear in the moments equation. You may also need to resolve forces in the vertical and horizontal.

The magnitude and/or direction of the reaction force at the hinge may be required. To evaluate the magnitude, Pythagoras' Theorem is needed, using the two components of the force. Finding the direction will require trigonometry.

Chapter 6
Momentum

6.1 Introduction

This chapter introduces the concept of **momentum**, sometimes referred to as **linear momentum**. An object's momentum is defined as the product of its mass and its velocity.

Momentum is conserved, which means that if a collision occurs between two objects, the total momentum of the two objects before the collision is equal to the total momentum after the collision. This is true whether the objects **coalesce** (stick together), or continue moving separately.

Modelling assumptions
Objects involved in collisions are modelled as particles in this chapter.

The conservation of momentum holds in an idealised setting in which there is no friction. Therefore, questions on this topic often state that the objects involved are moving on a smooth surface. If this information is not given in the question, you may state it if asked for any modelling assumptions that have been made.

Key words
- **Momentum**: The product of an object's mass and its velocity. The total momentum in a system is conserved during a collision, explosion, etc.
- **Particle**: A theoretical object that has a mass but no size.
- **Coalesce**: Stick together.

Before you start
You should:
- Recall your work with the constant acceleration formulae.
- Remember how to work with vector quantities in two dimensions.

Exercise 6A (Revision)

1. A particle P moves with a constant acceleration of 8 m s^{-2}. After P has travelled 27 m, it is moving with a velocity of 32 m s^{-1}. Find its initial velocity.

2. A car moves with constant velocity $(10\mathbf{i} + 4\mathbf{j})$ m s^{-1}. It starts at a point with position vector $(28\mathbf{i} - 17\mathbf{j})$ m. Find the position vector of the car after 10 seconds.

What you will learn
In this chapter you will learn how to:
- Calculate an object's momentum.
- Use the principle of conservation of momentum to solve problems involving direct collisions and explosions, problems involving coalescing objects and objects connected by string.
- Use these techniques in both one and two dimensions. In two dimensions this will involve using vectors with two components.

In the real world...
Deep beneath the France-Switzerland border, near to the city of Geneva, a mind-blowing physics experiment is taking place. Scientists working at the Large Hadron Collider (LHC) accelerate two beams of sub-atomic particles up to the speed of light and send them hurtling around a 27-kilometre circular tunnel in opposite directions. The two beams of particles then smash into each other. The collision of these high-energy particles is analysed in great detail, but the scientists have to be quick. The debris created by the collision decays within a fraction of a second, but contain some of the secrets of the universe. For example, sometimes there is evidence of strange new subatomic particles that may have been theorised, but not previously seen. One such particle, the Higgs Boson, is discussed further in the **In the Real World** section of chapter 11. The scientists

can also collect clues about other mysteries, such as the nature of dark matter, which appears to make up 27% of the universe. Since its discovery in the 1990s, theories have been put forward to explain dark matter, but its true nature remains an unsolved problem.

6.2 Calculation Of An Object's Momentum

An object's momentum is defined as the product of its mass and its velocity.

The units for momentum are kg m s^{-1}. You may also come across the equivalent units N s (newton seconds).

Throughout this chapter, when working with momentum in one dimension, we let left to right be the positive direction. This convention is widely used in other textbooks and exam material.

If an object is moving in the negative direction (right to left), it has a negative velocity and a negative momentum.

Worked Example

1. A skier has a mass of 75 kg and travels downhill at a speed of 25 m s^{-1}. Find the skier's momentum.

$$\text{Momentum} = mv$$
$$= 75 \times 25$$
$$= 1875 \text{ kg m s}^{-1}$$

When finding the total momentum in a system, add the momentum of each object. It is important to consider the direction of motion of each object, as the following example demonstrates.

Worked Example

2. Two balls roll towards each other on a smooth surface, as shown in the diagram. Ball A has a mass of 2 kg and a speed of 3 m s^{-1}. Ball A has a mass of 1 kg and a speed of 5 m s^{-1}.

3 m s^{-1} →   5 m s^{-1} ←

A 2 kg   B 1 kg

Find the total momentum of the balls.

Momentum of ball A = 2 × 3 = 6 kg m s^{-1}

Since ball B is moving from right to left it has a negative velocity.

Momentum of ball B = 1 × −5 = −5 kg m s^{-1}

Total momentum = 6 − 5 = 1 kg m s^{-1}

Momentum as a vector

Since velocity is a vector quantity, momentum is also a vector quantity.

Therefore, when working in two dimensions, an object's momentum can be written in component form.

Worked Example

3. A snooker ball of mass 100 grams is moving with a velocity of $(0.5\mathbf{i} - 3\mathbf{j})$ m s^{-1} across the snooker table. Find the snooker ball's momentum as a vector in component form.

$m = 0.1$ and $\mathbf{v} = 0.5\mathbf{i} - 3\mathbf{j}$

$$\text{Momentum} = m\mathbf{v}$$
$$= 0.1(0.5\mathbf{i} - 3\mathbf{j})$$
$$= (0.05\mathbf{i} - 0.3\mathbf{j}) \text{ kg m s}^{-1}$$

Exercise 6B

1. Which of these two objects has the greatest momentum? Show your working.
   - A car of mass 1500 kg travelling at 30 m s^{-1}; or
   - A ship of mass of 500 tonnes approaching a dock at a speed of 1 cm s^{-1}.

2. Two marbles roll directly towards each other, as shown in the diagram. The blue marble has a mass of 20 grams and a speed of 10 cm s^{-1}. The red marble has a mass of 25 grams and a speed of 16 cm s^{-1}.

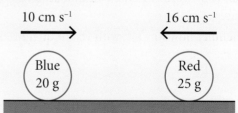

10 cm s^{-1} →   16 cm s^{-1} ←

Blue 20 g   Red 25 g

Find the combined momentum of the two marbles in kg m s^{-1}.

Exercise 6B...

3. A boy on a scooter has a mass of 45 kg and a velocity of $(16\mathbf{i} + 20\mathbf{j})$ m s^{-1}. Find the boy's momentum as a vector in component form.

4. Two particles are moving across a smooth surface. Particle P has a mass of 0.1 kg and a velocity of $(20\mathbf{i} - 30\mathbf{j})$ m s^{-1}. Particle Q has a mass of 0.2 kg and a velocity of $(-15\mathbf{i} + 35\mathbf{j})$ m s^{-1}. Find the combined momentum of the two particles.

5. A pea shot from a pea shooter has momentum of 0.02 kg m s^{-1}. If the pea travels with a velocity of 10 m s^{-1}, find its mass.

6. In the winter sport curling, players slide a stone across ice, aiming for a target. One stone has a mass of 19.1 kg and a momentum of $(\mathbf{i} + 45\mathbf{j})$ kg m s^{-1}. Find the stone's **speed**.

6.3 Conservation Of Momentum

According to the principle of **conservation of momentum**, the total momentum before a collision or explosion is equal to the total momentum afterwards. The conservation of momentum equation for two objects colliding is:

$$m_1 u_1 + m_2 u_2 = m_1 v_1 + m_2 v_2$$

where:

- m_1 and m_2 are the masses of the two objects;
- u_1 and u_2 are the velocities of the two objects before impact; and
- v_1 and v_2 are the velocities of the two objects after impact.

Collisions

In questions involving the conservation of momentum it is common to draw two diagrams: one before the collision (or explosion, etc) and one after. Example 4 below demonstrates this.

Worked Example

4. Two particles A and B of masses 1 kg and 2 kg respectively are moving towards each other along the same straight line with speeds 4 m s^{-1} and 2 m s^{-1} respectively. After impact the direction of motion of A is reversed and its speed is 2 m s^{-1}. Find the speed of B.

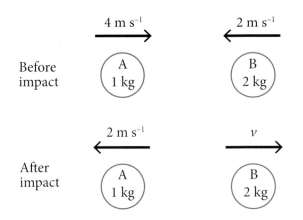

It is unclear in which direction B moves after the impact. The arrow has been drawn from left to right arbitrarily. Let its velocity after impact be v m s^{-1}.

By the principle of conservation of momentum:

$$m_1 u_1 + m_2 u_2 = m_1 v_1 + m_2 v_2$$
$$(1 \times 4) + (2 \times -2) = (1 \times -2) + 2v$$
$$4 - 4 = -2 + 2v$$
$$2v = 2$$
$$v = 1 \text{ m s}^{-1}$$

Since v is positive, B moves to the right with speed 1 m s^{-1}.

> **Note:** Be careful with the signs for all the velocities when using the conservation of momentum equation. In this example, particle B has a velocity of -2 m s^{-1} before the impact. Particle A changes direction, so its velocity is -2 m s^{-1} after the impact.

In the following example, the speeds of both objects after the impact are calculated.

Worked Example

5. A snooker ball P moving with a speed of 4 m s^{-1} hits a stationary ball Q of equal mass. After the impact both balls move in the same direction along the same straight line, but the speed of Q is twice that of P.
   (a) Find the speed of both balls after impact.
   (b) What modelling assumptions have been made in answering this question?

Let the balls have a mass of m kg and let the speed of P be v m s^{-1} after impact. Since the speed of Q is twice the speed of P, Q has a speed of $2v$ m s^{-1}.

(a)

Before impact

$$4 \text{ m s}^{-1} \qquad 0 \text{ m s}^{-1}$$

P
m kg

Q
m kg

After impact

$$v \text{ m s}^{-1} \qquad 2v \text{ m s}^{-1}$$

P
m kg

Q
m kg

Use the conservation of momentum equation:

$$m_1 u_1 + m_2 u_2 = m_1 v_1 + m_2 v_2$$
$$(m \times 4) + (m \times 0) = (m \times v) + (m \times 2v)$$
$$4m = mv + 2mv$$
$$4m = 3mv$$
$$v = \frac{4}{3} = 1.33 \text{ m s}^{-1} \text{ (3 s.f.)}$$

After impact P moves with speed 1.33 m s^{-1} and Q moves with speed of $2v$, which is $\frac{8}{3}$ or 2.67 m s^{-1} (3 s.f.)

(b) Both balls have been modelled as particles.

··

Exercise 6C

1. Two balls X and Y roll towards each other on a smooth table. Ball X has a mass of 3 kg and a speed of 4 m s^{-1}. Ball Y has a mass of 4 kg and a speed of 2 m s^{-1}. After colliding, X moves in the opposite direction with half its original speed. Find the speed and direction of Y after the collision.

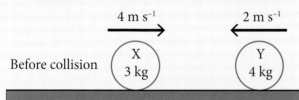

Before collision

$$4 \text{ m s}^{-1} \qquad 2 \text{ m s}^{-1}$$

X
3 kg

Y
4 kg

2. Two particles A and B move towards each other on a smooth surface, with A on the left and B on the right. Particles A and B have masses of 1 kg and 2 kg respectively. Before the particles collide, the speeds of A and B are 4 m s^{-1} and 3 m s^{-1} respectively. After the collision they both change their directions. Particle A now moves with three times the speed of B. Calculate the speeds of both balls after the collision and the directions in which they are both moving.

Exercise 6C...

3. Three smooth balls P, Q and R, with masses 7 kg, 3 kg and 1 kg respectively, can roll on a smooth surface in a straight line, as shown. Initially, P and Q have velocities 5 m s^{-1} and 2 m s^{-1} respectively. R is stationary.

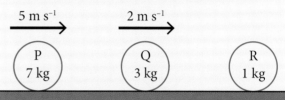

$$5 \text{ m s}^{-1} \qquad 2 \text{ m s}^{-1}$$

P
7 kg

Q
3 kg

R
1 kg

   (a) P collides with Q. After this collision, Q moves with a speed of 6.2 m s^{-1}. Find the velocity of P.
   (b) Subsequently, Q collides with R. Given that R starts moving with a velocity of 7.41 m s^{-1}, find the velocity of Q after this second collision.

4. Two blocks B and C slide across ice and collide directly. Blocks B and C have masses of M kg and 3 kg respectively. Before the collision they are moving towards each other with speeds of 3 m s^{-1} and 2 m s^{-1}. After the collision they both change direction. Block B now moves with speed 1.5 m s^{-1} and block C with speed 0.5 m s^{-1}, as shown in the diagram.

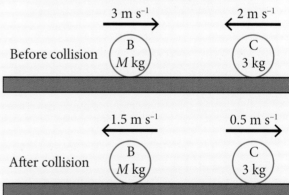

$$3 \text{ m s}^{-1} \qquad 2 \text{ m s}^{-1}$$

Before collision

B
M kg

C
3 kg

$$1.5 \text{ m s}^{-1} \qquad 0.5 \text{ m s}^{-1}$$

After collision

B
M kg

C
3 kg

   Find the value of M.

5. A small smooth sphere X of mass 4 kg is moving in a straight line with speed 3 m s^{-1} when it collides with another small smooth sphere Y of mass 2 kg, moving at 5.5 m s^{-1}, as shown in the diagram that follows. Immediately after the collision, X and Y are moving in opposite directions with speeds x m s^{-1} and y m s^{-1} respectively.

Exercise 6C...

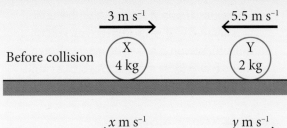

Before collision

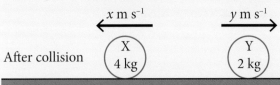

After collision

Given that $x : y = 3 : 7$, find the values of x and y.

Particles connected by a string

The principle of conservation of momentum also applies to two objects connected by a string.

Consider two particles P and Q, which are at rest on a smooth horizontal surface and are connected by a light inextensible string, which is initially slack.

Suppose Q is given a velocity in the positive direction (to the right).

After a short time, the string becomes taut. The particles then move with the same speed, since the string remains taut. This speed can be calculated by conservation of momentum.

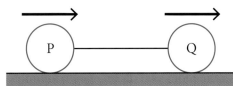

This is sometimes known as **jerk** in a string. At the moment the string becomes taut, Q experiences a jerk to the left and P a jerk of equal magnitude to the right.

Worked Example

6. Two particles P and Q of masses 3 kg and 6 kg respectively are connected by a light inextensible string. Initially they are at rest on a smooth table with the string slack. Q is projected directly away from P with a speed of 3 m s^{-1}. Find the common speed of the particles when the string becomes taut.

Particle P begins with a velocity of 0 m s^{-1}, while Q has a velocity of 3 m s^{-1}.

Before jerk

Let the common speed of P and Q after the jerk be v m s^{-1}.

After jerk

Using the principle of conservation of momentum:
$$m_1 u_1 + m_2 u_2 = m_1 v_1 + m_2 v_2$$
$$(3 \times 0) + (6 \times 3) = 3v + 6v$$
$$9v = 18$$
$$v = 2 \text{ m s}^{-1}$$

The common speed of the particles after the string goes taut is 2 m s^{-1}.

Exercise 6D

1. Two particles P and Q of mass 7 kg and 1 kg respectively are at rest on a smooth horizontal table. They are connected by a light inelastic string which is initially slack. Q is projected away from P with a speed of 4 m s^{-1}. Find the common speed of the particles after the string becomes taut.

2. Ball A of mass 1 kg is at rest 0.2 m from the edge of a smooth horizontal table. It is connected by a string of length 0.7 m to a particle B of mass 0.5 kg. Ball B is initially at rest at the edge of the table, as shown.

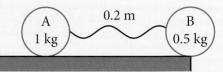

(a) Ball B then falls off the table. Assuming ball B has an initial speed of 0 m s^{-1} and that it does not reach the ground, find:
(i) the speed of ball B when the string becomes taut;

Exercise 6D...

(ii) the speed with which A begins to move.
(b) What modelling assumptions have been made in answering part (a)?

3. Two blocks A and B, with masses of 50 kg and 30 kg respectively, are connected by a rope, as shown. Block A is projected **towards** block B with a speed of 6 m s⁻¹.

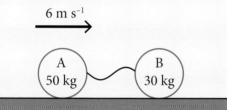

After colliding with block B, block A continues to move to the right. Block B also moves to the right, 2 m s⁻¹ faster than block A.
(a) Form an equation and solve it to find the speeds of both blocks after the collision. You may assume the rope does not get in the way when A and B collide.
(b) Block B moves faster than block A and after a short time the rope becomes taut. At this point the blocks begin to move with the same speed. Find this speed.

4. Two smooth spheres P and Q, with masses 1 kg and 5 kg respectively, are connected by a slack string, as shown.

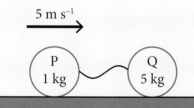

Sphere P is projected towards sphere Q with a speed of 5 m s⁻¹, as shown. Sphere Q is stationary. Sphere P collides with sphere Q.
(a) After the collision, sphere Q moves with a speed of ⅝ m s⁻¹. Find the speed and direction of sphere P.
(b) After a short time, the string becomes taut and the spheres begin to move with the same velocity. What is this common velocity?
(c) State any assumptions made in answering parts (a) and (b) of this question.

Exercise 6D...

5. Two particles A and B, of masses 3 kg and 4 kg respectively, are connected by a light, inextensible string. They are moving with velocities $(2\mathbf{i} - 3\mathbf{j})$ m s⁻¹ and $(-4\mathbf{i} + 6\mathbf{j})$ m s⁻¹ respectively. After a time the string becomes taut. Find the velocity with which the two particles begin to move together.

Explosions
In an explosion momentum is conserved.

Questions involving explosions usually involve two objects that separate due to the force of the explosion. The firing of a gun, for example, involves the gun and the bullet.

Since both objects usually start with zero velocity, the total momentum is initially zero. Since momentum is conserved, the total momentum remains zero after the explosion.

If the bullet moves with a positive velocity, the gun must gain a negative velocity, i.e. it moves in the opposite direction.

The person firing the gun experiences this as a **recoil**, in which the gun is forced back into their hand or shoulder as it fires.

Worked Example
7. A gun of mass 1.5 kg fires a bullet of mass 30 grams. The bullet leaves the gun with a speed of 300 m s⁻¹. Find the recoil speed of the gun.

The bullet's mass is 30 grams, which is 0.03 kg.

Take the direction of travel for the bullet to be the "positive" direction, i.e. its velocity is 300 m s⁻¹.

$$m_1u_1 + m_2u_2 = m_1v_1 + m_2v_2$$
$$1.5(0) + 0.03(0) = 1.5v_1 + 0.03(300)$$
$$0 = 1.5v_1 + 9$$
$$v_1 = -\frac{9}{1.5} = -6 \text{ m s}^{-1}$$

This means that the gun moves in the opposite direction to the bullet with a speed of 6 m s⁻¹.

In the following example, the velocity of the single object is non-zero before the explosion.

Worked Example

8. A shell of mass $6M$ kg is travelling horizontally, due east, with a speed of $5U$ m s^{-1}. After a short time, the shell explodes, breaking into two parts, A and B, of masses $2M$ and $4M$ kg respectively. A and B continue to move along the same horizontal straight line. After the explosion, the direction of B has not changed, but its speed is now $10U$ m s^{-1}. Find, in terms of U, the speed of A after the explosion and state its direction.

The total momentum before the explosion is
$6M \times 5U = 30MU$

Therefore, the total momentum afterwards is also $30MU$.

Let the velocity of A after the explosion be V. Then:
$(2M \times V) + (4M \times 10U) = 30MU$
$$2MV + 40MU = 30MU$$
$$2MV = -10MU$$
$$V = -5U \text{ m s}^{-1}$$

So part A is now travelling to the west with a speed of $5U$ m s^{-1}.

Exercise 6E

1. Particles A and B with masses 100 grams and 300 grams are initially in contact, as shown in the diagram. An explosion causes them to separate. Particle B moves to the right with speed 15 m s^{-1}. Find the speed and direction of particle A after the explosion.

Before explosion

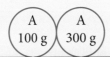

After explosion

2. A bullet of mass 50 grams is fired from a gun of mass 2 kg. The bullet leaves the gun with a speed of 300 m s^{-1}. Find the recoil speed of the gun.

Exercise 6E...

3. A TV explodes. The glass screen of mass 10 kg hits the sofa with a speed of 15 m s^{-1}. The plastic casing, of mass 5 kg, travels in the opposite direction, hitting the window.
   (a) Find the speed with which the casing hits the window.
   (b) What assumptions and approximations have been made in answering this question?

4. A satellite, which is in a stationary orbit above the Earth's surface, breaks into two sections. One section, with a mass of 350 kg, begins to move with a velocity of $(-26\mathbf{i} + 39\mathbf{j})$ m s^{-1}. The second section has a mass of 650 kg. Find its velocity.

5. **You may use, as an approximation, $g = 10$ m s^{-2} in this question.**
   A firework of mass 50 grams is launched vertically upwards from ground level with a speed of 20 m s^{-1}.
   (a) When it has reached a height of 10.2 m, the firework is still moving upwards with a speed of v m s^{-1}. Show that $v = 14$. You may assume that the firework moves freely under gravity from the moment of launch until it reaches this height.
   (b) At the height of 10.2 m, the firework explodes into two parts. One part, with a mass of 30 grams, continues to travel vertically upwards with a speed of 25 m s^{-1}. Find the speed and direction of the second part of the firework immediately after the explosion.

Coalescing objects

When objects collide and **coalesce**, they stick together and move as a single object.

The conservation of momentum equation for two objects colliding and coalescing is:

$$m_1 u_1 + m_2 u_2 = (m_1 + m_2)v$$

where v is the velocity of the combined object after the impact.

The following example involves three objects. In the second part, two of the objects collide and coalesce.

Worked Example

9. The diagram shows three particles A, B and C, with masses $3m$, $2m$ and $4m$ kg respectively, moving in the same straight line on a smooth surface.

(a) Particles A and B are moving towards each other with speeds u and $2u$ m s^{-1} respectively, as shown in the second diagram.

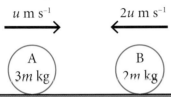

Particles A and B collide directly. Immediately after the collision particle B travels at u m s^{-1} and has reversed its direction. Find, in terms of u, the velocity of particle A after the collision.

(b) Particle B then collides with particle C, which is moving towards B at u m s^{-1}, as shown in the third diagram.

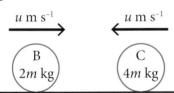

When particles B and C collide, they coalesce.
(i) Find in terms of u, the **speed** of the combined particle.
(ii) State whether the combined particle collides with particle A, justifying your answer.

(a) The conservation of momentum equation for particles A and B is:

$$m_A u_A + m_B u_B = m_A v_A + m_B v_B$$
$$(3m \times u) + (2m \times -2u) = (3m \times v_A) + (2m \times u)$$
$$3mu - 4mu = 3mv_A + 2mu$$
$$3mv_A = -3mu$$
$$v_A = -u$$

After the collision particle A moves with a velocity of $-u$ m s^{-1}.

> **Note:** This means particle A moves to the left with a speed of u m s^{-1}.

(b) (i) The conservation of momentum equation for the coalescing particles B and C is:

$$m_B u_B + m_C u_C = (m_B + m_C)v$$
$$(2m \times u) + (4m \times -u) = (2m + 4m)v$$
$$2mu - 4mu = 6mv$$
$$6mv = -2mu$$
$$v = -\frac{1}{3}u$$

The combined particle moves with a velocity of $-\frac{1}{3}u$ m s^{-1} after the collision (to the left with a speed of $\frac{1}{3}u$ m s^{-1}).

(ii) The combined particle will not collide with particle A. They are both moving to the left, but A is moving faster.

Exercise 6F

1. A smooth sphere A of mass 2.5 kg lies at rest on a smooth horizontal table. A second smooth sphere B of mass 1.5 kg is moving with speed 4 m s^{-1} and collides directly with A. The two spheres coalesce. Find the speed of the combined object after impact.

2. Objects A and B, with masses 1 kg and 0.5 kg respectively collide and coalesce. Before the collision, A has a velocity of 3 m s^{-1} and B is stationary. Find the velocity of the combined object after the collision.

3. A railway truck is travelling at 8 m s^{-1} along a straight piece of track when it hits an identical truck, which is stationary. The two trucks couple together.
   (a) Find the common speed of the trucks as they move together after the collision.
   (b) What modelling assumptions have you made in answering part (a)?

4. Three smooth balls A, B and C of equal size lie at rest in a straight line on a smooth horizontal table, as shown in the diagram.

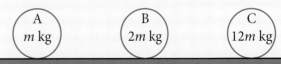

   Ball A has mass m kg.
   Ball B has mass $2m$ kg.
   Ball C has mass $12m$ kg.

Exercise 6F...

(a) Ball A is projected towards ball B with speed u m s^{-1}. Ball A is brought to rest by the collision. Find the speed of B after the collision.

(b) Ball B then collides with ball C. The speed of C after this collision is $\frac{u}{9}$ m s^{-1}.

Find the velocity of B after the collision.

(c) Explain briefly why a subsequent collision between balls A and B will occur.

(d) When ball B collides with ball A again, the two balls coalesce. Find their combined speed and direction.

5. Two asteroids X and Y, with masses 10 gigatonnes and 40 gigatonnes respectively, collide. Asteroid X has an initial velocity of $(4\mathbf{i} - 3\mathbf{j})$ km s^{-1}. Asteroid Y can be modelled as being initially stationary.

(a) If the asteroids combine and move together as a single object, find the speed with which it moves after the collision.

(b) Suppose instead the impact causes the combined object to break into two parts of equal mass. One part moves with a velocity of $(0.4\mathbf{i} - 0.1\mathbf{j})$ m s^{-1}. Find the velocity of the second part.

6.4 Summary

The momentum of an object is given by the product of its mass and its velocity:

Momentum $= mv$

Using vector notation for motion in two dimensions:

Momentum $= m\mathbf{v}$

In ideal collisions, explosions and other interactions between objects, momentum is conserved. The total momentum before the event is equal to the momentum afterwards.

Therefore, the conservation of momentum equation is:

$$m_1 u_1 + m_2 u_2 = m_1 v_1 + m_2 v_2$$

Objects that coalesce during a collision move together as a single object after the impact. In this situation, the conservation of momentum equation becomes:

$$m_1 u_1 + m_2 u_2 = (m_1 + m_2)v$$

Two objects connected by a string also move together as a single object if the string becomes taut, that is the two objects share a common velocity.

Chapter 7
Impulse

7.1 Introduction

This chapter introduces the concept of **impulse**. Impulse is a measure of the change in momentum given to an object during a collision or explosion.

The impulse given to an object by a force can be calculated as the magnitude of the force multiplied by the time for which the force acts.

Impulse, like momentum, is a vector quantity. Therefore, when working in two dimensions it can be written using component form.

Key words

- **Momentum**: the product of an object's mass and its velocity. The total momentum in a system is conserved during a collision, explosion, etc.
- **Particle**: a theoretical object that has a mass but no size.
- **Coalesce**: Stick together.
- **Impulse**: The impulse given to an object during a collision or explosion is its change in momentum.

Before you start

You should:

- Understand the concept of momentum, which is the product of an object's mass and its velocity.
- Understand that momentum is conserved in collisions and explosions.
- Know how to work with the conservation of momentum equation.

Exercise 7A (Revision)

1. Find the momentum of each of these objects.
   (a) A squash ball with a mass of 25 grams, travelling at 45 m s^{-1}.
   (b) A speedboat carrying 2 people, travelling at 18 m s^{-1}. The boat has a mass of 2000 kg and each person has a mass of 75 kg.

2. Two particles X and Y are moving in the same direction along a straight line on a smooth horizontal surface. The particles X and Y have masses pm kg and m kg respectively where p is a positive constant and $p > 1$. Initially the velocities of X and Y are 4 m s^{-1} and 2 m s^{-1} respectively, as shown in the diagram below.

Particle X collides with particle Y. After the collision, X continues to move in the same direction but now with a velocity of 3 m s^{-1}.

(a) Find an expression in terms of p for the velocity of Y after the collision.
(b) If Y is now travelling at twice its original velocity, find the value of p.

What you will learn

In this chapter you will learn:

- That the impulse given to an object is equal to the object's change in momentum.
- How to calculate the impulse given to an object during a collision or explosion.
- How to use the relationship: impulse = force × time
- How to work with impulse as a vector quantity.

In the real world...

What is the longest goal ever scored in football history and who holds the record?

In the women's game, goals scored from a player's own half are rare, but the record for the longest goal scored in the men's game now belongs to goalkeeper Tom King. He scored from 105 yards (96 metres) for Newport County against Cheltenham Town in a League Two match played on January 19, 2021. The Guinness Book of World Records subsequently confirmed the record.

Clearly Tom King's right foot gave the football a large **impulse** for it to travel such a distance. But this is not the only factor in his goal entering the record books. The vertical angle at which the ball was launched was important too, as well as the horizontal direction. And, although King may not admit it, there may have been a fair amount of luck involved!

Impulse is an important quantity in most ball sports. Top tennis players, for example, must return a tennis ball travelling towards them at well over one hundred miles per hour. The impulse the tennis racket transfers to the ball is huge, and this happens hundreds of times in a single match. This requires huge power and stamina on the part of top tennis players – they really have to be top athletes too!

7.2 Impulse In One Dimension

The impulse given to an object during a collision is its change in momentum. Since its momentum after the collision is mv and the momentum before is mu,

Impulse $= mv - mu$

In an ideal collision, the impulse given to one object is of equal magnitude but opposite sign to the impulse given to the other object.

The units for impulse are usually given as newton seconds (N s), although these units are equivalent to the units for momentum, kg m s^{-1}.

Worked Example

1. A ball of mass 0.5 kg bounces off a wall. As it approaches the wall it has a speed of 3 m s^{-1}. After impact, the ball rebounds with a speed of 1.5 m s^{-1}. Find the magnitude of the impulse given to the ball during the impact.

The ball begins with a velocity of 3 m s^{-1}, so $u = 3$.

The ball rebounds with a speed of 1.5 m s^{-1}. Since this is in the opposite direction to its original motion, it now has a **velocity** of -1.5 m s^{-1}, so $v = -1.5$.

Impulse $= mv - mu$

$\qquad = m(v - u)$

$\qquad = 0.5(-1.5 - 3)$

$\qquad = -2.25$

The **magnitude** of the impulse is 2.25 N s.

If a constant force F acts on a particle for time t, then the impulse provided by the force is of magnitude Ft.

Worked Example

2. An ice hockey player's stick hits the puck, which has a mass of 150 grams. The puck is initially at rest. After the impact, the puck moves with a speed of 12 m s^{-1}.
    (a) Calculate the magnitude of the impulse given to the puck as it is hit.
    (b) The stick is in contact with the puck for 0.1 s. Assuming the stick applies a constant force to the puck during this time, find the size of this force.

(a) Impulse $=$ change in momentum

$\qquad\qquad = mv - mu$

$\qquad\qquad = 0.15(12 - 0)$

$\qquad\qquad = 1.8$ N s

The puck has been given an impulse of 1.8 N s

(b) $\qquad Ft = 1.8$

$\qquad F \times 0.1 = 1.8$

$\qquad\qquad F = \dfrac{1.8}{0.1} = 18$ N

The following example demonstrates that the impulses given to the two objects during a collision are of equal magnitude, but have opposite signs.

Worked Example

3. Ball X, moving with velocity 5 m s^{-1} approaches ball Y, which is moving in the same direction with velocity v m s^{-1}, as shown in the diagram.
    The balls collide when ball X catches up with ball Y.

    After the collision, ball X continues to move in the same direction with a speed of v m s^{-1}. Ball Y continues to move in the same direction with a speed of 4 m s^{-1}.

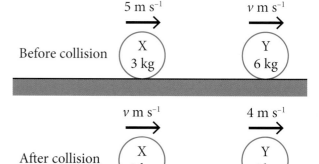

(a) Find v.
(b) Find the magnitude of the impulse given to ball Y by ball X during the collision.
(c) Write down the magnitude of the impulse given to ball X by ball Y.

(d) The impact lasts 0.3 seconds. Find the average force that ball X exerts upon ball Y during the impact.

(e) What modelling assumptions have been made in answering this question?

(a) By the conservation of momentum:
$$m_1 u_1 + m_2 u_2 = m_1 v_1 + m_2 v_2$$
$$(3 \times 5) + 6v = 3v + (6 \times 4)$$
$$15 + 6v = 3v + 24$$
$$3v = 9$$
$$v = 3 \text{ m s}^{-1}$$

(b) The impulse given to ball Y is ball Y's change in momentum.

For ball Y, the change in momentum is
$$m(v - u) = 6(4 - 3) = 6 \text{ kg m s}^{-1}$$

Therefore, the magnitude of the impulse given to ball Y is 6 N s.

> **Note:** Momentum is usually given in the units kg m s^{-1} and impulse in the units N s, although these are exactly equivalent.

(c) The impulse given to ball X is –6 N s. The **magnitude** of the impulse is 6 N s.

> **Note:** The impulse given to ball X is the same magnitude but a different sign to the impulse given to ball Y. As an exercise, the reader can check this by calculating the change in momentum for ball X.

(d) Impulse is the force applied multiplied by the time the force is acting. So, if the impulse is 6 N s, then:
$$Ft = 6$$
$$F \times 0.3 = 6$$
$$F = \frac{6}{0.3} = 20 \text{ N}$$

(e) The balls are modelled as particles, the surface is smooth.

Some problems may require the use of Newton's second law $F = ma$, and possibly the constant acceleration formulae, in conjunction with impulse calculations. It is important to remember that, when using both formulae $F = ma$ and Impulse = Ft, F represents the resultant force acting. Therefore, in cases of vertical motion, remember to include an object's weight force in force calculations.

Worked Example

4. Take $g = 10$ m s^{-2} as an approximation for the acceleration due to gravity in this question. Jim is a juggler. He is performing with three balls, each of mass 100 grams. When a ball lands in his right hand, Jim must throw it back up in the air almost instantly. One of Jim's balls lands in his hand with a downwards speed of 3 m s^{-1}. Jim throws it upwards with the same speed.

(a) Find the magnitude of the impulse Jim gives to the ball as he changes its direction.

(b) Given that Jim's hand is in contact with the ball for 0.2 s:

   (i) Show that the average force he exerts upon the ball during that time is 4 N.

   (ii) Find the acceleration of the ball during that time.

(a) Take upwards as the positive direction. Then the ball's initial velocity is $u = -3$ m s^{-1} and its velocity after the change in direction $v = 3$ m s^{-1}.
$$\text{Impulse} = m(v - u) = 0.1(3 - -3) = 0.6 \text{ N s}$$

(b) (i) Impulse = Ft
$$0.6 = F \times 0.2$$
$$F = \frac{0.6}{0.2} = 3$$

This is the resultant force acting on the ball. Let the force Jim applies to the ball be F_J newtons. Then: 
$$F_J - mg = 3$$
$$F_J = 3 + mg$$
$$= 3 + 0.1 \times 10$$
$$= 4 \text{ N}$$

   (ii)
$$v = u + at$$
$$3 = -3 + 0.2a$$
$$0.2a = 6$$
$$a = \frac{6}{0.2} = 30 \text{ m s}^{-2}$$

Exercise 7B

1. A ball of mass 400 grams bounces on the ground. Before it bounces it has a speed of 3 m s^{-1}. It bounces upwards with half the speed of impact. Find the magnitude of the impulse given to the ball by the ground as it bounces.

2. Ball B, moving with velocity 8 m s^{-1} approaches ball C, which is moving in the same direction with velocity 1 m s^{-1}, as shown in the diagram. The balls collide when ball B catches up with ball C. After the collision, ball B moves in the same direction with a speed of v m s^{-1}. Ball C also moves in the same direction with a speed of $(v + 2)$ m s^{-1}.

Exercise 7B...

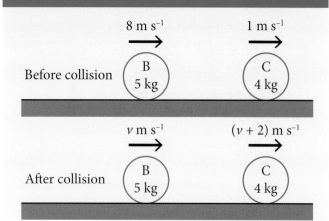

Before collision

8 m s⁻¹

B
5 kg

1 m s⁻¹

C
4 kg

After collision

v m s⁻¹

B
5 kg

$(v + 2)$ m s⁻¹

C
4 kg

(a) Find v.
(b) Find the magnitude of the impulse given to ball C by ball B during the collision.
(c) Write down the magnitude of the impulse given to ball B by ball C.
(d) During the impact, the average force that ball B exerts upon ball C is 100 N. Find how long the impact takes in seconds.
(e) What modelling assumptions have been made in answering this question?

3. Two particles P and Q of masses 3 kg and 2 kg respectively are at rest on a smooth horizontal table. They are connected by a light inelastic string which is initially slack. P is projected away from Q with a speed of 10 m s⁻¹. Find:
   (a) The common speed of the particles after the string becomes taut.
   (b) The magnitude of the impulse given to particle Q as the string becomes taut.

4. A small smooth sphere A of mass 0.2 kg is moving in a straight line with speed 4 m s⁻¹ when it collides with another small smooth sphere B of mass 0.8 kg, which is at rest. Immediately after the collision, A and B are moving in opposite directions with speeds x m s⁻¹ and y m s⁻¹ respectively. Given that $x:y = 2:3$, calculate:
   (a) The value of x.
   (b) The magnitude of the impulse exerted on A by B.

5. A bullet of mass 10 grams is fired from a gun of mass 1.5 kg. The gun recoils with a speed of 90 cm s⁻¹.
   (a) Find the speed of the bullet.

Exercise 7B...

(b) The bullet hits a rubber target and sinks 2.5 cm into the rubber. Find the bullet's deceleration.
(c) Find the average resistive force that the bullet experiences as it decelerates.
(d) Find the magnitude of the impulse transferred to the bullet as it decelerates.

6. A small sphere S of mass 0.08 kg moves with speed 1.5 m s⁻¹. It collides directly with another small sphere T, of mass 0.12 kg, which is moving in the same direction with a speed of 1 m s⁻¹. Immediately after the collision, S and T continue to move in the same direction with speeds U m s⁻¹ and V m s⁻¹ respectively. Given that $U:V = 21:26$,
   (a) Show that $V = 1.3$
   (b) Find the magnitude of the impulse received by S as a result of the collision.

7. Two particles X and Y are travelling directly towards each other on a smooth horizontal surface. Particle X has a mass of 3 kg and a speed of 4 m s⁻¹, as shown in the diagram.

Before impact

4 m s⁻¹

X
3 kg

Y

Particles X and Y collide. In the collision the magnitude of the impulse given by X to Y is 15 N s.
   (a) Find the velocity of particle X after the collision.
   (b) Particle Y has mass m kg and its speed before the collision is 4 m s⁻¹. After the collision the speed of particle Y is 1 m s⁻¹.
   (c) Find the two possible values of m.

8. A bullet of mass 0.04 kg is travelling horizontally at 400 m s⁻¹ when it hits a vertical target. The speed of the bullet is reduced to 150 m s⁻¹, 0.005 seconds after the bullet hits the target.
   (a) Find the change in momentum of the bullet.
   (b) Find the resistive force exerted by the target on the bullet.

Exercise 7B...

9. Take, as an approximation to the acceleration due to gravity, $g = 10$ m s^{-2} in this question. A pile driver of mass 225 kg is used to drive a steel post of mass 45 kg into the ground. The pile driver is released from rest at a height of 7.2 metres vertically above the post.
   (a) Show that the speed with which the pile driver hits the post is 12 m s^{-1}.
   (b) The pile driver and post move together immediately after the impact. Find the speed with which they move.
   (c) The pile driver and post come to rest 0.1 seconds after the impact. Find the resistance force from the ground.

10. In a shooting competition, a competitor fires a bullet from a rifle with a speed of 270 m s^{-1}. The mass of the rifle is 4.5 kg and the mass of the bullet is 0.15 kg.
   (a) Find the speed of recoil of the gun on the competitor's shoulder.
   (b) The rifle comes to rest after 0.05 seconds. Find the average force exerted by the gun on the competitor's shoulder during this time.

7.3 Impulse As A Vector Quantity

Since velocity is a vector quantity, impulse is also a vector quantity.

Therefore, when working in two dimensions, the impulse given to an object can be written as a vector in component form.

Impulse is the change in an object's momentum, so using vectors:

   Impulse $= m\mathbf{v} - m\mathbf{u}$

If a force $\mathbf{F}$ newtons acts on an object for t seconds, then the impulse given to that object is given by:

   Impulse $= \mathbf{F}t$

Worked Example

5. A force $\mathbf{F} = (-10\mathbf{i} + 15\mathbf{j})$ N acts on a particle for 25 seconds.
   (a) Find the impulse given to the particle during this time.
   (b) During this time, the particle accelerates from a velocity of $(19\mathbf{i} - 12\mathbf{j})$ m s^{-1} to $\mathbf{v}$ m s^{-1}. Given that the particle has a mass of 5 kg, find $\mathbf{v}$.

(a) Impulse $= \mathbf{F}t$
$$= (-10\mathbf{i} + 15\mathbf{j}) \times 25$$
$$= (-250\mathbf{i} + 375\mathbf{j}) \text{ N s}$$

(b) The particle has an initial velocity of
   $\mathbf{u} = (19\mathbf{i} - 12\mathbf{j})$ m s^{-1}

   Impulse $= m\mathbf{v} - m\mathbf{u} = m(\mathbf{v} - \mathbf{u})$
   $$(-250\mathbf{i} + 375\mathbf{j}) = 5(\mathbf{v} - (19\mathbf{i} - 12\mathbf{j}))$$

   Divide by 5:
   $$(-50\mathbf{i} + 75\mathbf{j}) = \mathbf{v} - (19\mathbf{i} - 12\mathbf{j})$$
   $$\mathbf{v} = (-50\mathbf{i} + 75\mathbf{j}) + (19\mathbf{i} - 12\mathbf{j})$$
   $$\mathbf{v} = (-31\mathbf{i} + 63\mathbf{j}) \text{ m s}^{-1}$$

In the following example two spheres collide. Their velocities are given as vectors.

Worked Example

6. Two spheres S and T have masses of 3 kg and 2 kg respectively. S is moving with a velocity of $(5\mathbf{i} + 3\mathbf{j})$ m s^{-1} and T with a velocity of $(4\mathbf{i} + 1.5\mathbf{j})$ m s^{-1} when they collide, as shown in the diagram.

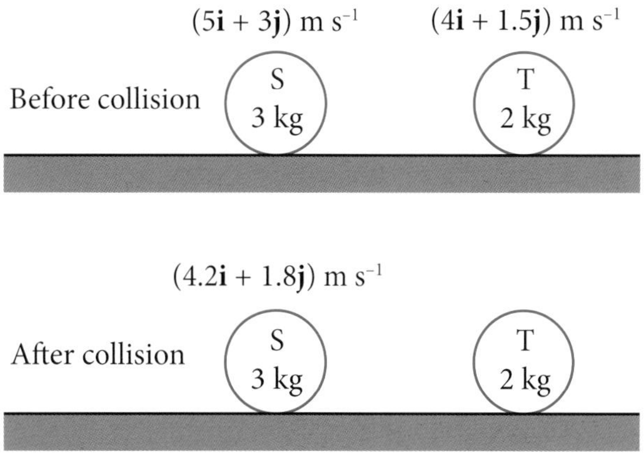

After the collision, S has a velocity of $(4.2\mathbf{i} + 1.8\mathbf{j})$ m s^{-1}.
(a) Find the velocity of T.
(b) Find, as a vector, the impulse given to sphere T during the collision.

(a) The conservation of momentum equation for the two spheres is:
$$m_S\mathbf{u}_S + m_T\mathbf{u}_T = m_S\mathbf{v}_S + m_T\mathbf{v}_T$$
$$3(5\mathbf{i} + 3\mathbf{j}) + 2(4\mathbf{i} + 1.5\mathbf{j}) = 3(4.2\mathbf{i} + 1.8\mathbf{j}) + 2\mathbf{v}_T$$
$$23\mathbf{i} + 12\mathbf{j} = 12.6\mathbf{i} + 5.4\mathbf{j} + 2\mathbf{v}_T$$
$$2\mathbf{v}_T = 10.4\mathbf{i} + 6.6\mathbf{j}$$
$$\mathbf{v}_T = (5.2\mathbf{i} + 3.3\mathbf{j}) \text{ m s}^{-1}$$

(b) The impulse given to sphere T is its change in momentum:

Impulse $= m(\mathbf{v} - \mathbf{u})$

$$= 2\big((5.2\mathbf{i} + 3.3\mathbf{j}) - (4\mathbf{i} + 1.5\mathbf{j})\big)$$

$$= (2.4\mathbf{i} + 3.6\mathbf{j})\,\text{N s}$$

....

The next example and some of the questions in the following exercise draw on your learning from chapter 1 (Kinematics), in conjunction with the work on momentum and impulse.

In the example, the object's variable acceleration vector is calculated using $\mathbf{F} = m\mathbf{a}$. Then calculus is required to obtain the velocity vector. The third part involves impulse calculations.

....

Worked Example

7. A particle P of mass 0.5 kg moves under the action of a single force $\mathbf{F}$ newtons. At time t seconds, $\mathbf{F} = (1.5t^2 - 3)\mathbf{i} + 2t\mathbf{j}$. When $t = 2$, the velocity of P is $(-4\mathbf{i} + 5\mathbf{j})\,\text{m s}^{-1}$.
    (a) Find the acceleration of P at time t seconds.
    (b) Show that, when $t = 3$, the velocity of P is $(9\mathbf{i} + 15\mathbf{j})\,\text{m s}^{-1}$.
    (c) When $t = 3$, the force is removed. Simultaneously, the particle experiences an impulse of magnitude 26 N s in the direction of the vector $5\mathbf{i} + 12\mathbf{j}$. Find the subsequent velocity of P.

(a) $\mathbf{F} = (1.5t^2 - 3)\mathbf{i} + 2t\mathbf{j}$ and $m = 0.5$

$$\mathbf{F} = m\mathbf{a}$$

$$(1.5t^2 - 3)\mathbf{i} + 2t\mathbf{j} = 0.5\mathbf{a}$$

$$\mathbf{a} = \big((3t^2 - 6)\mathbf{i} + 4t\mathbf{j}\big)\,\text{m s}^{-2}$$

(b) The acceleration is variable, since it depends on t. Therefore, the constant acceleration formulae cannot be used. Instead we require the techniques covered in chapter 1 (Kinematics) with variable acceleration:

$$\mathbf{a} = (3t^2 - 6)\mathbf{i} + 4t\mathbf{j}$$

$$\mathbf{v} = \int \mathbf{a}\,dt$$

$$\mathbf{v} = \int (3t^2 - 6)\mathbf{i} + 4t\mathbf{j}\,dt$$

$$\mathbf{v} = (t^3 - 6t)\mathbf{i} + 2t^2\mathbf{j} + \mathbf{c}$$

When $t = 2$, $\mathbf{v} = (-4\mathbf{i} + 5\mathbf{j})$

$$\therefore -4\mathbf{i} + 5\mathbf{j} = \big(2^3 - 6(2)\big)\mathbf{i} + 2(2)^2\mathbf{j} + \mathbf{c}$$

$$\therefore -4\mathbf{i} + 5\mathbf{j} = -4\mathbf{i} + 8\mathbf{j} + \mathbf{c}$$

$$\mathbf{c} = -3\mathbf{j}$$

$$\therefore \mathbf{v} = (t^3 - 6t)\mathbf{i} + 2t^2\mathbf{j} - 3\mathbf{j}$$

When $t = 3$:

$$\mathbf{v} = \big(3^3 - 6(3)\big)\mathbf{i} + 2(3)^2\mathbf{j} - 3\mathbf{j}$$

$$\mathbf{v} = (9\mathbf{i} + 15\mathbf{j})\,\text{m s}^{-1}$$

(c) The vector $5\mathbf{i} + 12\mathbf{j}$ has a magnitude of 13.

Therefore, the vector $10\mathbf{i} + 24\mathbf{j}$ has a magnitude of 26.

Therefore, a force of magnitude 26 N in the direction $5\mathbf{i} + 12\mathbf{j}$ is $10\mathbf{i} + 24\mathbf{j}$ N.

$$\mathbf{u} = 9\mathbf{i} + 15\mathbf{j}$$
$$m = 0.5$$

So:

$$m(\mathbf{v} - \mathbf{u}) = 10\mathbf{i} + 24\mathbf{j}$$

$$\mathbf{v} - \mathbf{u} = 20\mathbf{i} + 48\mathbf{j}$$

$$\mathbf{v} - (9\mathbf{i} + 15\mathbf{j}) = 20\mathbf{i} + 48\mathbf{j}$$

$$\mathbf{v} = (29\mathbf{i} + 63\mathbf{j})\,\text{m s}^{-1}$$

....

Exercise 7C

1. A force $\mathbf{F} = (-6\mathbf{i} + 9\mathbf{j})$ N acts on a particle for 15 seconds.
    (a) Find the impulse given to the particle during this time.
    (b) During this time, the particle accelerates to a velocity of $(\mathbf{i} + 6\mathbf{j})\,\text{m s}^{-1}$. Given that the particle has a mass of 3 kg, find its initial velocity.

2. Two particles Q and R, with masses 5 kg and 3 kg respectively are moving towards each other and collide directly. Before the collision, Particle Q moves with velocity $(2.5\mathbf{i} - 6.5\mathbf{j})\,\text{m s}^{-1}$ and R with velocity $(-7.5\mathbf{i} + 19.5\mathbf{j})\,\text{m s}^{-1}$. After the collision, Q moves with velocity $(-5\mathbf{i} - 6.5\mathbf{j})\,\text{m s}^{-1}$, as shown in the diagram.

(a) Find the velocity of particle R after the collision.

Exercise 7C...

(b) Find, as a vector, the impulse given to particle R during the collision.

3. Two particles X and Y have masses 2 kg and 1 kg respectively. They move in the same direction, particle X with velocity $(2\mathbf{i} + 3\mathbf{j})$ m s^{-1} and Y with velocity $(\mathbf{i} + 1.5\mathbf{j})$ m s^{-1}. Particle X catches up with particle Y and there is a collision. After the collision, X moves with velocity $(p\mathbf{i} + 2\mathbf{j})$ m s^{-1} and Y moves with velocity $\left(\frac{7}{3}\mathbf{i} + q\mathbf{j}\right)$ m s^{-1}, as shown in the diagram.

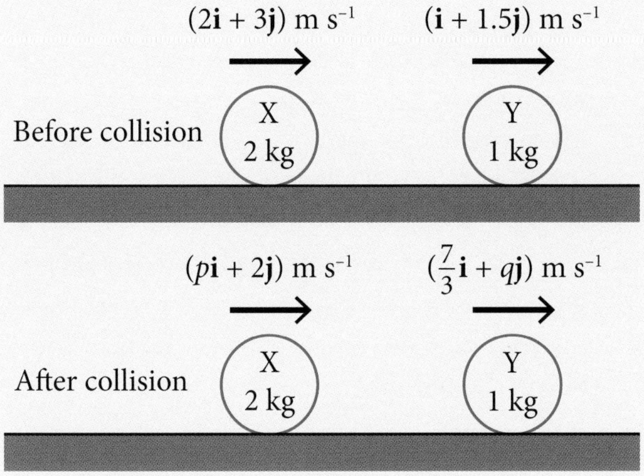

$(2\mathbf{i} + 3\mathbf{j})$ m s^{-1} $(\mathbf{i} + 1.5\mathbf{j})$ m s^{-1}

Before collision   X 2 kg   Y 1 kg

$(p\mathbf{i} + 2\mathbf{j})$ m s^{-1} $(\frac{7}{3}\mathbf{i} + q\mathbf{j})$ m s^{-1}

After collision   X 2 kg   Y 1 kg

(a) Show that $q = 3.5$ and find the value of p.
(b) Find, as a vector, the impulse given to particle Y during the collision.

4. A particle P of mass 0.4 kg is moving so that its position vector $\mathbf{r}$ metres at time t seconds is given by $\mathbf{r} = (t^2 + 4t)\mathbf{i} + (3t - t^3)\mathbf{j}$
(a) Calculate the speed of P when $t = 3$.

When $t = 3$ the particle P is given an impulse $(8\mathbf{i} - 12\mathbf{j})$ N s.
(b) Find the velocity of P immediately after the impulse.

5. At time t seconds $(t \geq 0)$, a particle P has position vector $\mathbf{p}$ metres with respect to a fixed origin O, where $\mathbf{p} = (3t^2 - 6t + 4)\mathbf{i} + (3t^3 - 4t)\mathbf{j}$
Find:
(a) The velocity of P after t seconds.
(b) The value of t when P is moving parallel to the vector $\mathbf{i}$.
(c) When $t = 1$, the particle P receives an impulse of $(2\mathbf{i} - 6\mathbf{j})$ N s. Given that the mass of P is 0.5 kg, find the velocity of P immediately after the collision.

Exercise 7C...

6. A particle P of mass 0.5 kg is moving under the action of a single force $\mathbf{F}$ newtons. At time t seconds, $\mathbf{F} = (6t - 5)\mathbf{i} + (t^2 - 2t)\mathbf{j}$
The velocity of P at time t seconds is $\mathbf{v}$ m s^{-1}. When $t = 0$, $\mathbf{v} = (\mathbf{i} - 4\mathbf{j})$ m s^{-1}.
(a) Find $\mathbf{v}$ at time t seconds.
(b) When $t = 3$, the particle P receives an impulse $(-5\mathbf{i} + 12\mathbf{j})$ N s. Find the **speed** of P immediately after it receives the impulse.

7.4 Summary

The impulse given to a particle is equal to the particle's change in momentum.

$$\text{Impulse} = mv - mu$$

If a constant force of magnitude F acts on a particle for time t, then the impulse of the force can be calculated as Ft.

$$\text{Impulse} = Ft$$

Like momentum, impulse is a vector quantity. Therefore, when working in two dimensions it can be written using component form. The two equations above become

$$\text{Impulse} = m\mathbf{v} - m\mathbf{u}$$

$$\text{Impulse} = \mathbf{F}t$$

Momentum and impulse are both measured in kg m s^{-1} or newton seconds (N s).

Chapter 8
Conditional Probability

8.1 Introduction

The probability of an event can change depending on the outcome of a previous event. For example, the probability that you are late for work may depend on whether you catch a train. This type of dependence can be modelled using **conditional probability**.

Key words

- **Random variable**: A variable whose value is dependent on a random phenomenon.
- **Mutually exclusive**: Two events are mutually exclusive if they cannot both occur at the same time.
- **Exhaustive events**: Two or more events are exhaustive if at least one of the events must occur.
- **Statistical dependence and independence**: Two events are independent if the occurrence of one does not affect the probability of the other event occurring. Otherwise, they are dependent.
- **Venn diagram**: A diagram in which circles are used to represent the occurrence of events and overlapping circles represent two or more events occurring.
- **Tree diagram**: A diagram representing a sequence of events, with each branch representing the occurrence of an event.
- **Conditional probability**: The probability of one event occurring given that another does occur.

Before you start
You came across probability at GCSE level and should know about:

- Calculating probabilities involving selection of items, both with and without replacement.
- Probability tree diagrams.
- Listing possible outcomes.

From AS Applied Mathematics you should know:

- The meaning of the words **intersection** and **union** and the notation used for each.
- How to use the addition and multiplication laws.
- About the following concepts:
  – mutually exclusive events;
  – exhaustive events;
  – statistical dependence and independence.

- How to calculate combined probabilities of up to three events using tree diagrams, Venn diagrams and two-way tables.

Exercise 8A (Revision)

1. In a chess set there are 16 white pieces and 16 black pieces. Eight of the white pieces are pawns and eight of the black pieces are pawns. Bobby takes two pieces from the chess set at random. Find the probability that:
   (a) both pieces are white,
   (b) one of the pieces is a pawn,
   (c) all three pieces are white pawns.

2. Pair up each diagram (a) to (c) with the correct description below. One description relates to the **complement**, one to the **intersection** and one to the **union**.

   Description 1
   The shaded region shows the event **not** A.
   It is also called the **complement** of A.
   It represents the event that A does not occur.
   It is denoted A' or $\overline{A}$.

   Description 2
   The shaded region shows the event A **and** B.
   This event is also called the **intersection** of A and B.
   It represents the event that both A and B occur.
   It is denoted $A \cap B$.

   Description 3
   This shaded region shows the event A **or** B.
   This event is also called the **union** of A and B.
   It represents the event that either A or B occurs.
   It is denoted $A \cup B$.

   (a)
   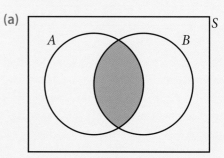

Exercise 8A...

(b)

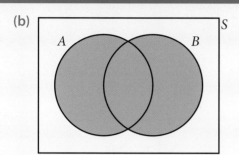

(c)

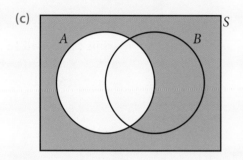

3. There are 100 beans in a packet of *Every Flavour Beans*. They can be **any** flavour, for example cherry, lemon, gravy, cabbage, washing up liquid, etc. Some are nice and some not so nice. Mia thinks that the red ones are most likely to be a fruit flavour. There are 20 red beans in the packet. There are 30 fruit-flavoured beans, some of which are red. There are 65 beans that are neither red nor fruit-flavoured.

   Mia chooses a bean at random. Let R be the event 'Mia chooses a red bean'. Let F be the event 'Mia chooses a fruit-flavoured bean'.
   (a) Find:
       (i) $P(R)$ **(ii)** $P(F)$ **(iii)** $P(R' \cap F')$
   (b) Find the probability that Mia chooses a bean that is both red and fruit-flavoured.

4. A and B are two events. $P(A) = 0.35$ and $P(B) = 0.15$ and $P(A \cap B) = 0.05$
   Find:
   (a) $P(A \cup B)$ **(b)** $P(B')$
   (c) $P(A \cap B')$ **(d)** $P(A \cup B')$

What you will learn

In this chapter you will learn:

- About conditional probability, including tree diagrams, Venn diagrams and two-way tables.
- About the conditional probability formula:
$$P(A|B) = \frac{P(A \cap B)}{P(B)}$$

In the real world...

Do you use email? Does your email provider give you a spam folder, where any junk email is automatically filed?

Most email services offer this and it's incredible how accurately junk mail is separated from the email you want to see. How does this work?

Algorithms are at work here. An algorithm is a decision-making piece of computer code. In the case of junk email, the algorithm uses a statistical method called Bayes Theorem. Bayes Theorem is beyond the scope of A-Level Applied Mathematics, but it depends heavily on conditional probability.

8.2 Conditional Probability

The probability of an event may depend upon a previous event. For example, the probability that you are involved in a car accident may depend on whether there is ice on the road. This type of dependence can be modelled using conditional probability.

Notation
You use a vertical line to indicate conditional probabilities. For example:

$P(B|A)$ means the probability that B occurs given that A has occurred.

$P(B|A')$ means the probability that B occurs given that A has not occurred.

> **Note:** If event B is independent of event A, then $P(B|A) = P(B|A') = P(B)$. The probability of B occurring does not depend on whether A has happened. This can be used to test for independence.

You can solve some problems involving conditional probability by considering a restricted sample space of the outcomes where one event has already occurred.

The following two examples demonstrate the use of restricted sample spaces.

Worked Examples

1. A school has 60 students in year 13. Of these, 17 study **only** arts subjects and 27 study **only** science subjects. Nine students study both arts and science subjects.

   A year 13 student is picked at random. Let *A* be the event 'the student studies at least one arts subject'. Let *S* be the event 'the student studies at least one science subject'.

   (a) Draw a two-way table to show this information.
   (b) Find:
       (i) $P(A' \cap S')$
       (ii) $P(S|A)$
       (iii) $P(A|S')$

(a)

	A	*A'*	Total
S	9	27	36
S'	17	7	24
Total	26	34	60

The cell containing 7 is calculated as
$60 - (17 + 27 + 9)$.

(b) The probabilities can be obtained from the table:

   (i) $P(A' \cap S') = \dfrac{7}{60}$

   (ii) $P(S|A) = \dfrac{9}{26}$

   |*A* means 'given the fact that the student chosen studies arts subjects'. Therefore, we are only interested in the numbers in the *A* column of the table. There are 9 students studying science subjects out of the 26 in this column.

   (iii) $P(A|S') = \dfrac{17}{24}$

   |*S'* means 'given the fact that the student chosen does not study science subjects'. Therefore, we are only interested in the *S'* row of the table. There are 17 students studying arts subjects out of the 24 in this row.

2. Two fair 6-sided dice are thrown, and the sum of the numbers shown is recorded.
   (a) Draw a sample space diagram showing the possible outcomes.
   (b) Given that the sum on the two dice is exactly 5, find the probability that at least one die lands on a four.

(a) The sample space diagram is shown below. Each number on the grid is the sum of the two numbers shown on the dice.

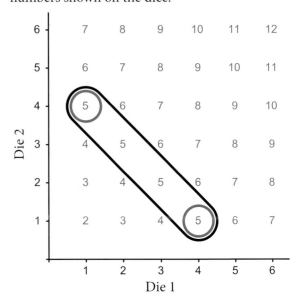

(b) The black line encloses all the ways in which a total of 5 can be achieved. This is the restricted sample space. There are 4 outcomes in the restricted sample space. The two grid points circled in blue represent one of the dice landing on a four.

Therefore $P(\text{one } 4 \mid \text{total is } 5) = \dfrac{2}{4} = \dfrac{1}{2}$

Exercise 8B

1. There are 50 students in Year 12 at Highway School. The two-way table shows the number of students taking school dinners and packed lunches.

	School dinner	Packed lunch	Total
Male	15	7	
Female	18	10	
Total			

   (a) Copy the table and fill in the total column on the right and the total row at the bottom of the table.
   (b) A student is chosen at random. Find:
       (i) $P(\text{Male})$
       (ii) $P(\text{Packed lunch} \mid \text{Male})$
       (iii) $P(\text{Male} \mid \text{Packed lunch})$
       (iv) $P(\text{School dinner} \mid \text{Female})$

Exercise 8B...

2. There are 120 members of a gym, of whom 65 are female and 55 male. Of the female members, 43 have annual membership, while the rest pay per visit. For the male members, 40 have annual membership with the rest paying per visit.
    (a) Draw a two-way table showing this information.
    (b) Find:
        (i) $P(\text{Male})$
        (ii) $P(\text{Annual membership} \mid \text{Male})$
        (iii) $P(\text{Female} \mid \text{Pay per visit})$

3. Two dice are thrown and the sum of the two numbers shown is recorded as S.
    (a) Draw a sample space diagram showing all the possibilities.
    (b) Using your sample space diagram, given that $S \geq 9$, find the probability that S is 10.

4. A blue and a green spinner both have five outcomes numbered 1 to 5. The two spinners are spun at the same time and the sum of the numbers shown is recorded as X.
    (a) Draw a sample space diagram for X.
    (b) Use your sample space diagram to help you find:
        (i) $P(X = 5)$
        (ii) $P(X = 9 \mid \text{Number on green spinner is 5})$
        (iii) $P(\text{Number on green spinner is 2} \mid X < 4)$
    (c) What modelling assumptions have you made when answering this question?

8.3 Venn Diagrams

You can find conditional probabilities from a Venn diagram by considering the section of the Venn diagram that corresponds to the restricted sample space.

> **Note:** Venn diagrams can be drawn with the **number** of items in each region of the diagram; or with the **probabilities** in each region. In Example 3 below, which relates to a 5-a-side football team, the Venn diagram has been drawn with the **number** of players in each region.

Worked Example

3. A university department's 5-a-side football team is put together. It comprises 5 people of ages 20, 20,

32, 50, 60. The two youngest people on the team are students. The events A and B are defined as follows:

A: A randomly selected player is a student.

B: The age of a randomly selected player is a multiple of 10.

(a) Draw a Venn diagram to show these two events.
(b) Use the Venn diagram to find:
    (i) $P(A)$ (ii) $P(B)$
    (iii) $P(A \cap B)$ (iv) $P(A \cup B)$
    (v) $P(A')$ (vi) $P(A|B)$
    (vii) $P(B|A)$

(a) Venn diagram:

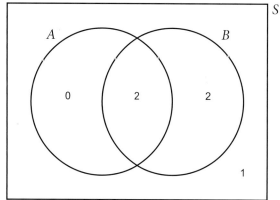

Note that the whole sample space is denoted by a rectangle, and that this is labelled 'S', or sometimes using a curly capital E, $\mathcal{E}$.

(b) (i) Two of the five players are in the A circle;

    Therefore $P(A) = \dfrac{2}{5}$

    (ii) Four of the five players are in the B circle;

    Therefore $P(B) = \dfrac{4}{5}$

    Note that these numbers include those in the intersection region.

    (iii) Two out of the five players are in the intersection region.

    Therefore $P(A \cap B) = \dfrac{2}{5}$

    (iv) Four out of the five players are inside one or both of the circles.

    Therefore $P(A \cup B) = \dfrac{4}{5}$

    (v) $P(A')$ means the probability of event A not taking place. It can be calculated as $1 - P(A)$:

    $$P(A') = 1 - P(A) = 1 - \frac{2}{5} = \frac{3}{5}$$

    This is sometimes called the **complement** of A.

(vi) To calculate $P(A|B)$ using the Venn diagram, consider the B circle as a restricted sample space. Find the probability of a player being in the A circle, given that the player is one of the four in the B circle. Then:

$$P(A|B) = \frac{2}{4} = \frac{1}{2}$$

(vii) To calculate $P(B|A)$, consider the A circle as a restricted sample space. Then:

$$P(B|A) = \frac{2}{2} = 1$$

Note: Later in this chapter you will learn the **conditional probability formula** to help with these calculations.

In the next example the Venn diagram has been drawn with probabilities in each of the regions.

Worked Example

4. A and B are two events such that $P(A) = 0.65$, $P(B) = 0.35$ and $P(A \cap B) = 0.2$.
   (a) Draw a Venn diagram showing this information.
   (b) Find:
   (i) $P(A|B)$ (ii) $P\big(B|(A \cup B)\big)$
   (iii) $P(A'|B')$
   (c) Determine whether events A and B are independent.

(a) Use the information given to complete the four regions of the diagram. It is easiest to complete the intersection region first, 0.2. Then, since the numbers in the A circle must add up to 0.65, calculate the left-hand region as being 0.45. The 0.15 is calculated in a similar way for the B circle. Finally, $0.45 + 0.2 + 0.15 = 0.8$, so the probability in the region outside of both circles is given by $1 - 0.8 = 0.2$. So the completed diagram is:

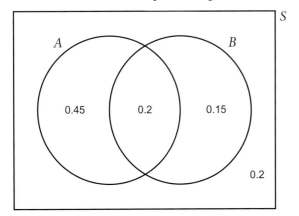

(b) (i) $P(A|B)$ can be calculated by considering event B as the restricted sample space.

$$P(A|B) = \frac{0.2}{0.2 + 0.15} = \frac{4}{7} = 0.571 \,(3\text{ s.f.})$$

(ii) $P\big(B|(A \cup B)\big)$ can be calculated by using $A \cup B$ as the restricted sample space. Then:

$$P\big(B|(A \cup B)\big) = \frac{0.2 + 0.15}{0.45 + 0.2 + 0.15} = \frac{7}{16}$$

(iii) To calculate $P(A'|B')$, consider B' as the restricted sample space, i.e. everything outside the B circle. The regions in question have probabilities of 0.45 and 0.2. Therefore:

$$P(A'|B') = \frac{0.2}{0.45 + 0.2} = \frac{4}{13}$$

(c) **Method 1**
Recall from AS Applied Mathematics the test for independence of two events A and B:
If $P(A) \times P(B) = P(A \cap B)$ then events A and B are independent.

$$P(A) = 0.65; P(B) = 0.35$$

$$P(A) \times P(B) = \frac{91}{400}$$

$$P(A \cap B) = 0.2$$

$$P(A) \times P(B) \neq P(A \cap B)$$

Therefore, events A and B are not independent.

Method 2
If events A and B are independent, then $P(A) = P(A|B) = P(A|B')$, i.e. the probability of event A happening does not depend on whether B has happened.

We were given $P(A) = 0.65$

In part (b)(i) we calculated $P(A|B) = 0.571 \,(3\text{ s.f.})$

Since $P(A) \neq P(A|B)$, events A and B are not independent.

Exercise 8C

1. Peadar supports two teams: Team A in the Northern Ireland Football League and Team B in Gaelic Football. He estimates that the probability of Team A winning is 0.4, the probability of Team B winning is 0.5 and the probability of both teams winning is 0.1.
   (a) Letting A be the event 'Team A wins' and B be the event 'Team B wins', draw a Venn diagram, showing the two events.
   (b) Find the probability that neither of Peadar's teams win.
   (c) Find the probability exactly one of Peadar's teams wins.
   (d) Given that Team A does not win, find the probability that Team B does win.

2. A letter is chosen at random from the alphabet.
   B is the event 'the letter appears in the word BELFAST'.
   L is the event 'the letter appears in the word LISBURN'.
   (a) Draw a Venn diagram, showing the events B and L and the associated **probabilities**.
   (b) Given that the letter chosen is in the word LISBURN, find the probability it is also in BELFAST?

3. A smartphone app has been developed to help people with sleeping problems. However, only 75% of people who download the app can get it to work. 85% of people who download it say they have sleeping problems. 65% of people who download the app have sleeping problems and can get the app to work.
   (a) Draw a Venn diagram to show this information.
   (b) A person who downloaded the app is chosen at random. Given that the person was able to get the app to work, find the probability they have sleeping problems. Give your answer as a fraction in its simplest form.

4. F and G are two events such that $P(F) = 0.8$, $P(G) = 0.4$ and $P(F \cap G) = 0.25$.
   (a) Draw a Venn diagram to represent this information.
   (b) Find:
   (i) $P(F \cup G)$ (ii) $P(F|G)$
   (iii) $P(G|F)$ (iv) $P(G'|F')$

Exercise 8C...

5. The Venn diagram shows the probabilities of 3 events, D, E and F.

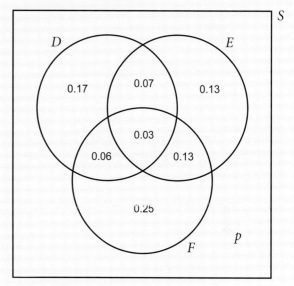

   (a) Find the missing probability p.
   (b) Find:
   (i) $P(D|E)$
   (ii) $P(F|D')$
   (iii) $P((D \cap E)|F')$
   (iv) $P(F|(D' \cup E'))$

6. The Venn diagram shows the number of students in a class who play hockey, football and netball.

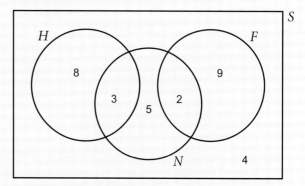

   One of these students is chosen at random.
   (a) Find the probability that the student plays:
   (i) football;
   (ii) exactly two of the sports.
   (b) Determine whether playing hockey and netball are independent events. Show your working.

Exercise 8C...

7. The Venn diagram shows the probabilities of events Q and R.

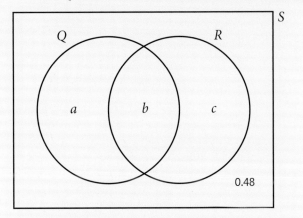

Given that Q and R are independent events and that $3b = 2c$, find the values of a, b and c.

8.4 Probability Formulae

The addition law – revision

In AS Applied Mathematics you learnt about the addition law for probabilities.

For two mutually exclusive events, A and B, there is no intersection of their circles on a Venn diagram. In this case $P(A \cup B) = P(A) + P(B)$

In cases where A and B are **not** mutually exclusive, there is an intersection between the circles, as shown in the Venn diagram below.

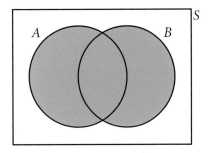

For events A and B that are not mutually exclusive,
$P(A \cup B) = P(A) + P(B) - P(A \cap B)$

This is known as the **addition law**.

Worked Example

5. Events A and B are not mutually exclusive. Given that $P(A) = 0.53$, $P(A \cup B) = 0.68$ and $P(A \cap B) = 0.05$, find $P(B)$.

Using the addition law:
$$P(A \cup B) = P(A) + P(B) - P(A \cap B)$$
$$0.68 = 0.53 + P(B) - 0.05$$
$$P(B) = 0.68 - 0.53 + 0.05 = 0.2$$

The multiplication law

In AS Applied Mathematics, you learnt that, for independent events:
$$P(A \cap B) = P(A) \times P(B)$$

This law can be extended to all pairs of events, independent or dependent:
$$P(A \cap B) = P(A) \times P(B|A)$$

When the multiplication law is extended in this way it sometimes referred to as the **conditional probability formula**. It can be rearranged to make $P(B|A)$ the subject:

$$P(B|A) = \frac{P(A \cap B)}{P(A)}$$

> **Note:** At the time of publication, both the addition law and the conditional probability formula appear in the Probability and Statistics section of the CCEA formula booklet.

In Example 3 earlier in this chapter, the ages of five footballers were presented in a Venn diagram. In the example that follows, this scenario is revisited to demonstrate the use of the conditional probability formula.

Worked Example

6. A university department's 5-a-side football team is put together. It comprises 5 people of ages 20, 20, 32, 50, 60. The two youngest people on the team are students. The events A and B are defined as follows:

   A: A randomly selected player is a student.

   B: The age of a randomly selected player is a multiple of 10.

   A Venn diagram showing the players' ages is as follows.

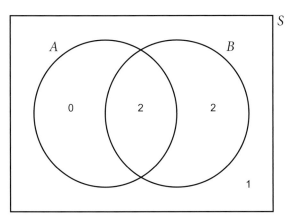

Find:
(a) $P(A)$ (b) $P(B)$
(c) $P(A \cap B)$ (d) $P(B|A)$
(e) $P(A|B)$

From the Venn diagram:

(a) $P(A) = \dfrac{2}{5}$

(b) $P(B) = \dfrac{4}{5}$

(c) $P(A \cap B) = \dfrac{2}{5}$

Using the conditional probability formula:

(d) $P(B|A) = \dfrac{P(A \cap B)}{P(A)} = \dfrac{2/5}{2/5} = 1$

(e) $P(A|B) = \dfrac{P(A \cap B)}{P(B)} = \dfrac{2/5}{4/5} = \dfrac{1}{2}$

Exercise 8D

1. A car dealer offers purchasers a three-year warranty on a new car. They sell two models, the Slowda and the Slugga. For the first 55 cars sold of each type the number of claims under the warranty is shown here.

Model	Claim	No claim
Slowda	44	11
Slugga	42	13

One of these purchases is chosen at random. Let A be the event that no claim is made by the purchaser under the warranty and let B be the event that the car purchased is a Slugga.
(a) Find $P(A \cap B)$
(b) Find $P(A')$

Exercise 8D...

(c) Given that the purchaser chosen does not make a claim under the warranty, use the conditional probability formula to find the probability that the car chosen is a Slowda.
(d) Show that making a claim is **not** independent of the make of the car purchased. Comment on this result.

2. A gym owner observes that if a married man and woman both join the gym, the probability that the man has a donor card is ½ and the probability that the woman has a donor card is ⅔. The probability that the man has a donor card given that the woman has a donor card is ⅑. A married man and woman are chosen at random.
(a) Use the multiplication law to show that the probability of them both having donor cards is ⁸⁄₂₇.
(b) Find the probability that only one of them has a donor card.
(c) Find the probability that neither of them has a donor card.

3. The events A and B are independent such that $P(A) = 0.18$ and $P(B) = 0.2$.
Find:
(a) $P(A \cap B)$
(b) $P(A \cup B)$
(c) $P(A|B')$

4. Three events A, B and C are defined such that A and B are mutually exclusive and A and C are independent. Given that $P(A) = 0.6$, $P(A \cup C) = 0.8$ and $P(B) = 0.3$ find:
(a) $P(A|C)$
(b) $P(A \cup B)$
(c) $P(C)$

5. For the events A and B, $P(A \cap B') = 0.31$, $P(A' \cap B) = 0.24$ and $P(A \cup B) = 0.71$
(a) Draw a Venn diagram to illustrate the complete sample space for the events A and B.
(b) Write down the values of $P(A)$ and $P(B)$.
(c) Find $P(A|B')$.
(d) Determine whether or not A and B are independent.

Exercise 8D...

6. A woman is trying to sleep. There is a 75% chance she gets to sleep before 3 a.m. There is a 25% chance there is a mosquito in her room. Given that there is a mosquito in her room, the probability of her getting to sleep before 3 a.m. is 5%. Find the probability there is a mosquito in her room **and** she gets to sleep before 3 a.m.

7. A university department specialises in entomology, the study of insects. Academics from the department have published 96 scientific papers.

   35 of the papers relate to ants.
   38 of them mention butterflies.
   44 of them relate to cockroaches.
   12 papers refer to both ants and butterflies.
   9 papers relate to both ants and cockroaches.
   16 papers relate to both butterflies and cockroaches.
   4 papers relate to all three species.

   (a) Draw a Venn diagram to show this information.

   One of the papers is chosen at random.
   (b) What is the probability the paper relates to cockroaches, but not ants or butterflies?
   (c) What is the probability the paper relates to butterflies or cockroaches or both?

   Given that the paper relates to butterflies or cockroaches:
   (d) Find the probability that it does not relate to ants.
   (e) Determine whether a paper relating to butterflies and a paper relating to cockroaches are independent events.

8.5 Tree Diagrams

Conditional probabilities can be represented on a tree diagram, as shown.

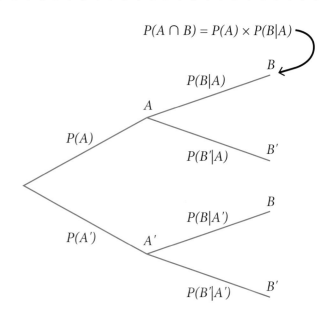

The probabilities on the second set of branches represent the conditional probabilities of B given that A has or has not happened.

....

Worked Example

7. Rachel has a little girl called Sophie and a dog called Cookie. The probability Rachel walks Sophie to school is 0.6. Cookie loves his walks. Sometimes he gets one walk, sometimes two.

   If Rachel walks Sophie to school, the probability Cookie gets two walks is 0.8. If Rachel does not walk Sophie to school, the probability Cookie gets two walks is 0.6.

   Given that Cookie gets two walks on Monday, find the probability Rachel walked Sophie to school.

   _____

   Let the two events "Rachel walks Sophie to school" and "Cookie gets two walks" be named S and C respectively. They can be represented on a tree diagram:

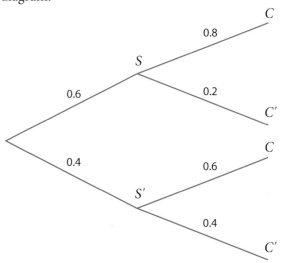

The word "given" in the question is a clue that conditional probability is required. We are being asked to find $P(S|C)$:

$$P(S|C) = \frac{P(S \cap C)}{P(C)}$$

From the tree diagram, $P(S \cap C) = 0.6 \times 0.8 = 0.48$

To find $P(C)$, it is important to use both routes through the tree diagram: S followed by C and S' followed by C.

$$P(C) = (0.6 \times 0.8) + (0.4 \times 0.6) = 0.72$$

$$P(S|C) = \frac{P(S \cap C)}{P(C)}$$

$$= \frac{0.48}{0.72} = \frac{2}{3}$$

Exercise 8E

1. A and B are two events such that $P(A) = 0.6$, $P(B|A) = 0.35$, and $P(B|A') = 0.25$.
   (a) Copy and complete the tree diagram representing this information.

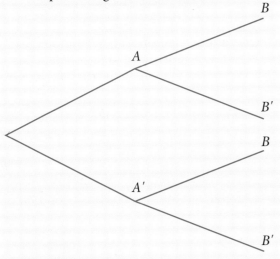

   (b) Find:
   (i) $P(A \cap B)$ (ii) $P(A' \cap B)$
   (iii) $P(B)$ (iv) $P(A|B)$

2. A bag contains 10 blue balls and 5 red balls. A ball is selected at random from the bag and its colour is recorded. The ball is not replaced. A second ball is selected and its colour is recorded.
   (a) Draw a tree diagram to represent this information.

Exercise 8E...

   (b) Find the probability that:
   (i) The second ball selected is red.
   (ii) Both balls selected are red, given that the second ball selected is red.

3. In a factory machines A, B and C all produce metal rods of the same length. Machine A produces 45% of the rods, machine B produces 30% and the rest are produced by machine C. Of their production of rods, machines A, B and C produce 2%, 4% and 7% defective rods respectively.
   (a) Draw a tree diagram to represent this information.
   (b) Find the probability that a randomly selected rod is produced by machine A and is defective.
   (c) Find the probability that a randomly selected rod is defective.
   (d) Given that a randomly selected rod is defective, find the probability that it was produced by machine C.

4. To get to work, Brídín first catches a bus and then catches a train.

   The probability that the bus is on time is 0.75
   If the bus is on time, then the probability that she catches the train is 0.85
   If the bus is late, then the probability that she catches the train is 0.55

   Given that Brídín catches the train, what is the probability that the bus was on time?

5. A certain type of cancer affects 1 in 100 men over the age of 60. In a doctor's surgery, all men over 60 are tested for this type of cancer. However, the tests are not perfect.

   About 2 in 100 men with the disease test negative (a false negative).
   About 3 in 100 men who don't have the disease test positive (a false positive).

   A man over 60 is tested for this type of cancer.
   (a) With the help of a probability tree diagram, find the probability he tests positive.
   (b) Given that he tests positive, show that there is roughly a 25% chance he has the disease.
   (c) Comment on your answer to part (b).

Exercise 8E...

6. Jill has a biased coin. It has a probability of p of landing on heads (H). Jill tosses the coin twice.
   (a) Find an expression in terms of p for the probability of getting HH.
   (b) Find an expression in terms of p for the probability of getting TT.
   (c) Find an expression in terms of p for the probability of getting either HH or TT.
   (d) Find an expression in terms of p for the probability of getting HH, given that both tosses of the coin produced the same outcome.
   (e) The probability that it was HH given that both outcomes were the same is 0.2. Show that $p = \frac{1}{3}$

8.6 Summary

The addition law is:

$$P(A \cup B) = P(A) + P(B) - P(A \cap B)$$

The multiplication law that you learnt in AS Applied Mathematics, for independent events, is:

$$P(A \cap B) = P(A) \times P(B)$$

This law can be extended to all pairs of events, independent or dependent:

$$P(A \cap B) = P(A) \times P(B|A)$$

This can be arranged to give the conditional probability formula:

$$P(B|A) = \frac{P(A \cap B)}{P(A)}$$

Two events A and B are independent if:

$$P(A) = P(A|B) = P(A|B')$$

Questions may involve drawing a Venn diagram or a tree diagram, or interpreting one of these.

Chapter 9
Normal Distribution

9.1 Introduction

The normal distribution is a continuous probability distribution with a characteristic bell-shaped curve. It can be used to model many real-life distributions, for example the heights of women in Northern Ireland, or the masses of carrots harvested from a field.

Notation

We use capital letters for random variables and lower case letters for particular values. For example, X could be the random variable 'the heights in centimetres of boys in Highway School'. We could ask questions such as 'Find $P(X > 150)$', or 'If $x = 150$, find $P(X > x)$'.

The letter a is also frequently used as an unknown data value, so you will see the notation $P(X > a)$.

If the random variable X is normally-distributed, with a mean of μ and a standard deviation of σ we write:

$$X \sim N(\mu, \sigma^2)$$

The convention is to state the mean and the variance inside the brackets. The variance is the square of the standard deviation σ.

Section 9.5 introduces a very important normally-distributed random variable Z, which has a mean of 0 and a standard deviation of 1. Particular values of Z are denoted by a lower case z. The distribution of Z is called the **standard normal distribution**. We write:

$$Z \sim N(0,1)$$

As a shorthand for $P(Z < z)$ we often use the notation $\Phi(z)$.

Key words

- **Normal distribution**: A probability distribution characterized by a bell-shaped curve. The normal distribution occurs a lot in natural processes.
- **Standard normal distribution**: A special case of the normal distribution, which has a mean of 0 and a standard deviation of 1.
- **Random variable**: A variable whose value depends on randomness in some way, for example the height of a randomly-chosen boy in a school.
- **Continuous**: A continuous variable is one that can take any value within a range. Examples are height, length and mass.

Before you start

You should know:

- How to calculate probabilities using a probability tree diagram.
- How to solve simultaneous equations in two unknowns.
- The meaning of the terms **mean** and **standard deviation** in a dataset.
- The meaning of the terms **upper quartile**, **lower quartile** and **interquartile range**.

Exercise 9A (Revision)

1. Solve the simultaneous equations:
   $$5a + 2b = 44$$
   $$7a - 3b = 50$$

2. On a farm, the probability that a newborn single lamb has a mass less than 5.5 kg is 0.1. Two new newborn single lambs are chosen at random. Using a probability tree diagram, or otherwise, find the probability that:
   (a) Both lambs have masses less than 5.5 kg.
   (b) Exactly one lamb has a mass less than 5.5 kg.

3. Cara carries out a survey of the speeds of vehicles on the A1 road near Newry.
   (a) On a Sunday, Cara finds that 25% of vehicles travel at 40 mph or less, while 25% travel at 65 mph or more. Find the interquartile range for the speeds on the Sunday.
   (b) The following day, Cara repeats the traffic survey at 8:30 in the morning. She finds that 25% of the vehicles are travelling at less than 30 mph, while the fastest 25% of vehicles are travelling at 45 mph or above. Find the interquartile range for the speeds on Monday.

Exercise 9A...

   (c) Cara also calculates the mean speed and the standard deviation of the speeds on both days.
- (i) Comment on how the mean speed changes.
- (ii) Comment on how the standard deviation changes.

   (d) Suggest a reason for the difference in the mean and standard deviation from one day to the next.

What you will learn

In this chapter you will learn:

- About the normal distribution as an example of a continuous probability distribution, and know the characteristics of the normal distribution curve.
- About the standard normal distribution.
- How to find probabilities using the normal distribution.
- How to find the range of values a normally-distributed variable can take, given a probability.
- How to find the mean and/or standard deviation, given information about a normally-distributed variable.

In the real world...

During World War II, bomber planes would come back from battle with bullet holes. The Allies initially sought to strengthen the most commonly-damaged parts of the planes to increase combat survivability.

However, mathematician Abraham Wald spotted that the pattern of bullet holes didn't quite match up with what would be expected using a standard probability distribution. He suggested that certain parts of the planes weren't covered in bullet holes because planes that were shot in those critical areas **did not** return.

This insight led to the planes' armour being reinforced on critical parts of returning planes where there were **no** bullet holes. This wisdom was also applied with some success to the American plane *Skyraider* during the Korean War.

Abraham Wald's work shows that, sometimes, missing data may be more meaningful than the available data itself. This is relevant to the work on missing data and data cleaning in AS Applied Mathematics, but Wald's work also demonstrated a very practical application of a theoretical probability distribution.

9.2 The Normal Distribution

In AS Applied Mathematics you learnt about the binomial distribution, which is an example of a **discrete probability distribution**. A **discrete random variable** can only take certain values, often integer values. For example, if ten coins are tossed, the discrete random variable X could be the number of heads.

The normal distribution is an example of a **continuous probability distribution**. A **continuous random variable** can take any one of infinitely many values. The probability that a continuous random variable takes any one specific value is 0. Instead, it is possible to discuss the probability that it takes values within a given range.

For example, in a field of sunflowers the continuous random variable X could be the height of a randomly chosen sunflower in centimetres. X can take any value within a range. You may be asked to find the probability the randomly chosen sunflower has a height between 50 and 60 cm, i.e. to find $P(50 \leq X \leq 60)$.

Continuous random variables in real life are, like the sunflower heights, likely to take values grouped around a central value, and less likely to take extreme values. The normal distribution is a continuous probability distribution that can be used to model many naturally occurring characteristics that behave in this way. Examples of continuous random variables that can be modelled using the normal distribution are:

- the heights of people within Northern Ireland;
- the masses of chimpanzees in a forest;
- the level of blood glucose in a certain population;
- the masses of breakfast cereal in boxes that are labelled '750 grams'.

For example, the heights of people within Northern Ireland may be modelled as having a normal distribution, with a mean of 165 cm and a standard deviation of 11 cm.

The normal distribution:

- has parameters μ, the population mean, and σ^2, the population variance;
- is a symmetrical distribution, meaning that the mean and the median are equal;
- has a bell-shaped curve with asymptotes at both ends;
- has a total area under the curve of 1.

The mean is the value of x that corresponds to the centre of the curve. It is the value that is associated with the greatest probability.

If X is normally-distributed, we write $X \sim N(\mu, \sigma^2)$. For example, if X is the random variable 'the height of a randomly chosen person in Northern Ireland', then

$$X \sim N(165, 11^2)$$

The distribution can be shown on a graph, as shown:

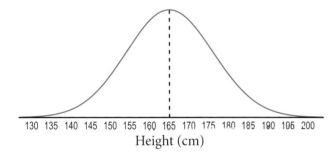

Height (cm)

Note the following points about the curve:

- If the y-axis is shown it should be labelled 'Probability density', but it is often omitted. The important aspects of the graph are the numbers on the x-axis and the area below the curve, which represents probability.

- Although a normal random variable could take any value, in practice it very rarely lies more than 5 standard deviations from the mean, so for x values at these distances from the mean, the probability density falls very close to zero.

- Approximately 68% of the data lies within one standard deviation of the mean.

- 95% of the data lies within two standard deviations of the mean.

- Nearly all the data (99.7%) lies within three standard deviations of the mean.

....

Worked Example

1. The diameter of corks produced in a bottling plant is a continuous random variable X mm, such that $X \sim N(10, 0.2^2)$. Find:
   (a) $P(X > 10)$
   (b) $P(9.8 < X < 10.2)$

(a) The distribution curve for the diameter of the corks is shown. On the diagram, the area representing $P(X > 10)$ is shown.

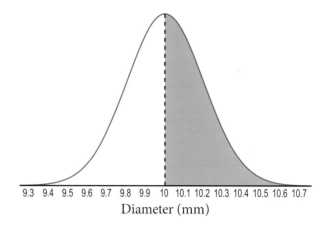

Diameter (mm)

Exactly half of the area under the curve is shaded due to the symmetry of the normal distribution curve. Since the entire area under the curve is 1:

$$P(X > 10) = 0.5$$

(b) The area corresponding to $P(9.8 < X < 10.2)$ is shown.

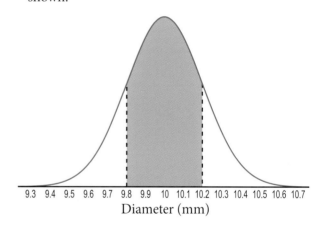

Diameter (mm)

The lower bound of 9.8 is $\mu - \sigma$ and the upper bound of 10.2 is $\mu + \sigma$.

Since we know that roughly 68% of corks have diameters within one standard deviation of the mean:

$$P(9.8 < X < 10.2) = 0.68$$

....

Exercise 9B

1. The diameter, D mm, of a sample of circular buttons being produced in a factory has a mean of 12.5 mm and a standard deviation of 0.1 mm. Sketch the probability distribution for D.

2. The masses, M grams, of a hundred guinea pigs are modelled as $M \sim N(\mu, 36)$. If 97.5% of the guinea pigs weigh less than 120 grams, find μ.
   Hint: Sketch the probability distribution for the

Exercise 9B...

guinea pigs. Use the fact that 95% of the masses lie within 2 standard deviations of the mean.

3. The heights, H metres, of trees in a forest are modelled as $H \sim N(\mu, \sigma^2)$. If 84% of the trees are less than 9 metres in height and 97.5% of the trees are more than 3 metres in height, find μ and σ.

   Hint: Sketch the probability distribution for the trees. Use the facts that 95% of the heights lie within 2 standard deviations of the mean and that 68% of the heights lie within 1 standard deviation of the mean. Then use simultaneous equations.

9.3 Finding Probabilities Using the Normal Distribution

You can find probabilities for the normal distribution using the normal cumulative distribution function on your calculator.

Worked Example

2. $Y \sim N(70, 8^2)$

   Find:

   (a) $P(Y < 80)$ (b) $P(Y \geq 60)$
   (c) $P(75 \leq Y \leq 80)$ (d) $P(Y < 60 \text{ or } Y > 85)$

   ___

   (a) $P(Y < 80)$

   A diagram helps you to visualise the question:

   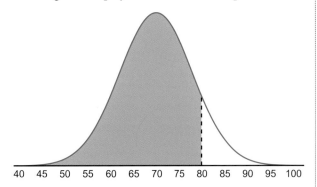

   Since more than half of the area under the curve is shaded, we are expecting an answer greater than 0.5.

   See section 9.7 for instructions on finding the required probability.

   $P(Y < 80) = 0.894$ (3 s.f.)

(b) $P(Y \geq 60)$

In a continuous distribution, there is no distinction between 'greater than' and 'greater than or equal to'. The area of interest under the curve is shown.

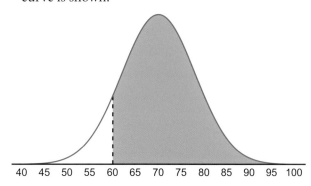

Again, we can see that an answer greater than 0.5 is expected.

$P(Y \geq 60) = 0.894$ (3 s.f.)

(c) To find $P(75 \leq Y \leq 80)$, again a sketch is helpful.

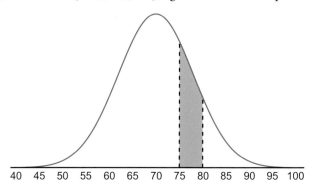

The sketch shows that we expect an answer much less than 0.5. From the calculator:

$P(75 \leq Y \leq 80) = 0.160$ (3 s.f.)

(d) To find $P(Y < 60 \text{ or } Y > 85)$:

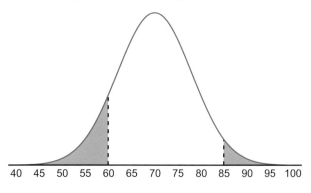

This probability relates to two distinct areas beneath the curve.

Method 1. From the calculator:
$P(Y < 60) = 0.10564 ...$
and $P(Y > 85) = 0.03039 ...$

$\therefore P(Y < 60 \text{ or } Y > 85) = 0.10564 ... + 0.03039 ...$
$= 0.136 \text{ (3 s.f.)}$

Method 2. From the calculator:
$P(60 < Y < 85) = 0.86395 ...$

Since the entire area under the curve is 1:
$P(Y < 60 \text{ or } Y > 85) = 1 - 0.86395 ...$
$= 0.136 \text{ (3 s.f.)}$

Exercise 9C

1. The random variable $X \sim N(50, 4^2)$
   Find:
   (a) $P(X < 55)$
   (b) $P(X > 40)$
   (c) $P(X \geq 20)$

2. The random variable $Y \sim N(20, 4)$
   Find:
   (a) $P(Y < 17)$
   (b) $P(Y > 26)$
   (c) $P(Y \leq 30)$
   (d) $P(18 < Y \leq 22)$

3. The random variable Q is normally-distributed
   such that $Q \sim N(85, 64)$
   (a) Find, to 4 decimal places,
   (i) $P(Q < 80)$
   (ii) $P(Q > 80)$
   (b) Calculate the sum of your answers from
   (a)(i) and (ii). Comment on your answer.

4. In July 2021, Northern Ireland's highest
   temperature of 31.3°C was recorded in
   Castlederg. On that day, the maximum
   temperature was recorded in several locations
   across Northern Ireland. The temperatures
   at these locations roughly followed a normal
   distribution with a mean of 26°C and a standard
   deviation of 2.5°C.
   (a) Estimate the probability that one of these
   temperature readings taken at random was
   (i) lower than 24°C,
   (ii) higher than 30°C.
   (b) Explain why your answers to part (a) are
   estimates.

Exercise 9C...

5. In a mountain range, the heights of the peaks
   can be modelled using a normal distribution
   with a mean of 1250 m and a standard deviation
   of 95 m. Tim sets himself a challenge to climb
   one of the mountains, chosen at random by his
   girlfriend.
   (a) Find the probability that Tim climbs
   a mountain with a height greater than
   1400 m.
   (b) Find the probability that Tim climbs a
   mountain whose height is greater than
   1000 m, but less than 1200 m.

6. A doctor carries out routine measurements
   of each patient's body mass index (BMI). A
   patient's BMI gives an indication of whether
   that patient is overweight. The doctor finds that
   BMI in her patient population can be modelled
   using a normal distribution with a mean of 23
   and a standard deviation of 4.5. If a patient's
   BMI is between 25 and 30 they are considered
   overweight. If a patient's BMI is above 30 they
   are considered obese.
   (a) Find, to three significant figures, the
   probability a patient chosen at random is:
   (i) overweight,
   (ii) obese.
   (b) The doctor puts 10% of her obese patients
   on a new medication. If the doctor has 3000
   patients in total, find how many patients
   receive the medication.

9.4 The Inverse Normal Distribution

For a given probability p, it is possible to use the
calculator to find a value of a such that $P(X < a) = p$. On
the calculator, this is usually called the **inverse normal
distribution** function.

Worked Examples

3. $X \sim N(40, 3^2)$
   Find, correct to 3 significant figures, the value of a,
   such that:
   (a) $P(X < a) = 0.6$
   (b) $P(X > a) = 0.25$
   (c) $P(35 < X < a) = 0.3$

See section 9.7 for instructions on how to use the
inverse normal distribution on the calculator.

(a) Sketch the distribution, showing $P(X < a) = 0.6$.

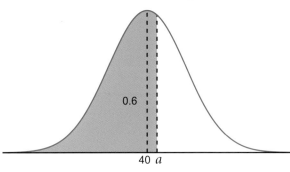

From the sketch, $a > 40$.
Using the inverse normal function on the calculator with Area = 0.6, $\sigma = 3$ and $\mu = 40$, we get:

xInv = 40.76004128

So $a = 40.8$ (3 s.f.)

(b) $P(X > a) = 0.25$

Since the entire area under the curve is 1, $P(X < a) = 0.75$:

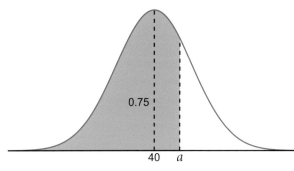

Using the inverse normal function on the calculator with Area = 0.75, $\sigma = 3$ and $\mu = 40$, we get:

xInv = 42.02346874

So $a = 42.0$ (3 s.f.)

(c) Draw a sketch for $P(35 < X < a) = 0.3$.

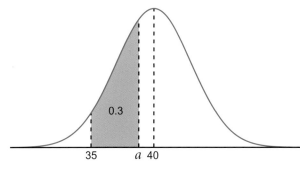

$P(35 < X < a) = 0.3 \Rightarrow P(X < a) - P(X < 35) = 0.3$

Using normal cumulative distribution on the calculator with $\sigma = 3$ and $\mu = 40$, $P(X < 35) = 0.04779 \ldots$ so:

$P(X < a) - 0.04779 = 0.3$
$\qquad P(X < a) = 0.3 + 0.04779 = 0.34779$

Using inverse normal the calculator gives

xInv = 38.82611836

So $a = 38.8$ (3 s.f.)

4. Packets of spaghetti are labelled as having a mass M of 200 grams. A sample shows that the masses are normally-distributed such that $M \sim N(205, 5^2)$
   (a) Given that 70% of the packets weigh less than m grams, find m.
   (b) Find the interquartile range for the masses.

   (a) 70% of packets weigh less than m grams. Therefore, if a packet is chosen at random, the probability of it having a mass of less than m grams is 0.7. So $P(M < m) = 0.7$

   Using the inverse normal function on the calculator with Area = 0.7, $\sigma = 5$ and $\mu = 205$, we get:

   xInv = 207.6220022

   So $m = 208$ grams (3 s.f.)

   (b) The interquartile range (IQR) is the difference between the upper quartile and the lower quartile, i.e. $Q_3 - Q_1$
   The upper quartile Q_3 is the value of M that 75% of the masses lie below. The lower quartile Q_1 is the value of M that 25% of the masses lie below.
   $P(M < Q_3) = 0.75$

   Using the inverse normal function:
   $Q_3 = 208.3724 \ldots$ and $Q_1 = 201.6275 \ldots$

   $\text{IQR} = Q_3 - Q_1$
   $\qquad = 208.3724 - 201.6275$
   $\qquad = 6.74$ (3 s.f.)

Exercise 9D

1. $X \sim N(60, 4^2)$
   Find, correct to 3 significant figures, the value of a, such that:
   (a) $P(X < a) = 0.7$
   (b) $P(X > a) = 0.3$
   (c) $P(55 < X < a) = 0.3$

Exercise 9D...

2. $Y \sim N(10, 0.5^2)$
   Find, correct to 3 significant figures, the value of a, such that:
   (a) $P(Y < a) = 0.35$
   (b) $P(Y > a) = 0.4$
   (c) $P(10.5 \leq Y \leq a) = 0.1$

3. Libby the chicken farmer takes a sample of 200 eggs produced on her farm in one month. She discovers that the eggs have masses X that roughly follow a normal distribution $X \sim N(80, 6^2)$. Libby chooses one of the eggs at random. Find the value of a to 3 significant figures, given that:
   (a) $P(X < a) = 0.72$
   (b) $P(X \geq a) = 0.998$
   (c) $P(a \leq X \leq 75) = 0.15$

4. The length of time T minutes spent by patients in a hospital department's waiting area can be modelled using a normal distribution, such that $T \sim N(20, 3^2)$. Jeremy enters the waiting area at 10:00 am. At what time is there is a 50% chance he will still be there?

5. Two hundred drivers take a test each month, hoping to qualify to drive a truck. In October the percentage scores S of the group is approximately normally-distributed as $S \sim N(63, 7.5^2)$
   (a) 65.5% of the drivers pass the test. What is the pass mark?
   (b) The top 11.5% of the drivers were given a merit for excellent driving. Find the lowest percentage score required for a merit.

6. The distribution of the waist size of men coming into a clothing shop is modelled using $W \sim N(36, 4^2)$. The shop stocks trousers with waist measurements, in inches, of 28, 30, 32, 34, 36, 38 and 40. The shop conducts a survey to determine how many pairs of each size of trousers they should stock.

   A man whose waist measurement lies between 27 and 29 inches is classified as being a size 28. A man whose waist measurement lies between 29 and 31 inches is classified as being a size 30. And so on.

   There is a 40% chance a customer is classified as being a size s or above. Find s.

Exercise 9D...

7. In a timber yard, the lengths L cm of the pieces of timber can be modelled using a normal distribution, such that $L \sim N(190, 15^2)$
   (a) Explain why the median value $Q_2 = 190$ cm
   (b) Find the interquartile range for the pieces of timber.

8. Pots of yogurt have a mean weight of 100 grams and a standard deviation of 15 grams. A pot of yogurt is chosen at random.
   (a) Find, to 3 significant figures, the probability that:
      (i) the pot weighs less than 120 grams;
      (ii) the pot weighs less than 80 grams;
      (iii) the pot weighs between 110 grams and 120 grams;
      (iv) the pot weighs more than 130 grams.
   (b) Three quarters of the pots of yogurt weigh less than w grams. Find w to one decimal place.

9.5   The Standard Normal Distribution

The **standard normal distribution** is sketched below. It has a mean of 0 and a standard deviation of 1.

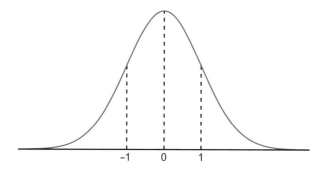

Transforming data

Suppose we have a random variable X which is normally-distributed with a mean of μ and a standard deviation of σ, so $X \sim N(\mu, \sigma^2)$

It is possible to transform individual x values into z values using this formula, or coding:

$$z = \frac{x - \mu}{\sigma}$$

The resulting z values are normally-distributed with a mean of 0 and a standard deviation of 1.

Notation

The standard normal distribution is written as

$$Z \sim N(0, 1^2)$$

For the standard normal distribution, $P(Z < a)$ is often written as $\Phi(a)$.

Worked Example

5. (a) Sketch the standard normal distribution and show, as an area, $\Phi(0.9)$
    (b) Find $\Phi(0.9)$ using your calculator.

(a) The sketch of the standard normal distribution is shown below. The area shaded represents $\Phi(0.9)$.

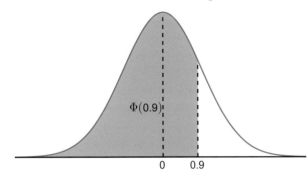

(b) To find $P(Z < 0.9)$ or $\Phi(0.9)$ using the calculator, use the normal cumulative distribution with these parameters:

Lower = –1000 (or any large negative number)
Upper = 0.9
$$\sigma = 1$$
$$\mu = 0$$

The calculator gives:
P = 0.8159398747

So $\Phi(0.9) = 0.816$ (3 s.f.)

Important results

We can derive some important results about the standard normal distribution from the symmetry of the curve.

Result 1:

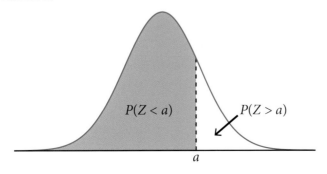

The entire area below the curve is 1, so:
$$P(Z > a) + P(Z < a) = 1$$
or
$$\mathbf{P(Z > a) = 1 - P(Z < a)}$$

Result 2:

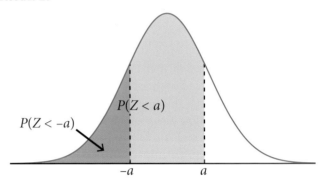

From the diagram, $P(Z < a) + P(Z < -a) = 1$, so:
$$P(Z < -a) = 1 - P(Z < a)$$
or
$$\mathbf{\Phi(-a) = 1 - \Phi(a)}$$

Result 3:

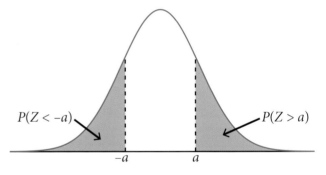

Because of the symmetry of the normal distribution,
$$\mathbf{P(Z < -a) = P(Z > a)}$$
Likewise:
$$\mathbf{P(Z > -a) = P(Z < a)}$$

In your formula booklet you have a table for the standard normal distribution, which is an alternative to using the calculator. Note that this table provides $P(Z < z)$, or $\Phi(z)$, for $z \geq 0$. The corresponding probabilities in the table are all greater than or equal to 0.5.

If you need to use the table to find $P(Z > z)$ or $P(Z < z)$ for $z < 0$, one of the results above is required.

Worked Example

6. The lengths of 100 lizards living on an island are measured. The lengths are found to be normally-distributed with a mean of 15 cm and a standard deviation of 2.4 cm. A lizard is chosen at random. Using the standard normal distribution, find the probability that the lizard's length is:
    (a) less than 18.6 cm;
    (b) less than 12 cm;
    (c) greater than 18 cm;
    (d) greater than 9 cm.

Let X be the random variable representing the lengths of this population of lizards. Then $X \sim N(15, 2.4^2)$.

To transform an x value into a z value, use $z = \dfrac{x - \mu}{\sigma}$

(a) If $x = 18.6$, then $z = \dfrac{18.6 - 15}{2.4} = 1.5$

$\therefore P(X < 18.6) = P(Z < 1.5) = \Phi(1.5)$
$$= 0.9332$$
$$= 0.933 \text{ (3 s.f.)}$$

(b) If $x = 12$, then $z = \dfrac{12 - 15}{2.4} = -1.25$

$P(X < 12) = P(Z < -1.25)$
$$= \Phi(-1.25)$$

Since $\Phi(-a) = 1 - \Phi(a)$ we can write:
$$= 1 - \Phi(1.25)$$

Using the calculator or the table in the formula booklet to find $\Phi(1.25)$:
$$= 1 - 0.89435 \ldots$$
$$= 0.106 \text{ (3 s.f.)}$$

(c) If $x = 18$, then $z = \dfrac{18 - 15}{2.4} = 1.25$

$P(X > 18) = P(Z > 1.25)$

Since $P(Z > a) = 1 - P(Z < a)$ we can write:
$$= 1 - P(Z < 1.25)$$
$$= 1 - 0.89435 \ldots$$
$$= 0.106 \text{ (3 s.f.)}$$

(d) If $x = 9$, then $z = \dfrac{9 - 15}{2.4} = -2.5$

$P(X > 9) = P(Z > -2.5)$

Since $P(Z > a) = 1 - P(Z < a)$ we can write:
$$= P(Z < 2.5)$$

Using the calculator or the table in the formula booklet to find $\Phi(2.5)$:
$$= 0.99379 \ldots$$
$$= 0.994 \text{ (3 s.f.)}$$

Note: All parts of this question can be done using the normal cumulative distribution function on the calculator. For example, in part (d), use:

Lower = 9, Upper = 1000 (or any large number), $\sigma = 2.4$, $\mu = 15$

The calculator gives $P = 0.9937903346$, so $P(X > 9) = 0.994$ (3 s.f.)

In this way use of the standard normal distribution and the table can be avoided.

However, it is important to understand the standard normal distribution as you may be specifically tested on this.

Exercise 9E

1. For the standard normal distribution $Z \sim N(0, 1^2)$, find the following to 4 decimal places. For some parts of this question you may use the fact that $P(Z > a) = 1 - P(Z < a)$
    (a) $P(Z < 2.1)$
    (b) $P(Z < 1.37)$
    (c) $P(Z > 1.6)$
    (d) $P(Z > 0.75)$

2. For the standard normal distribution $Z \sim N(0, 1^2)$, find the following to 4 decimal places. Use the fact that $\Phi(-a) = 1 - \Phi(a)$
    (a) $P(Z < -2.4)$
    (b) $P(Z < -1.58)$
    (c) $P(Z < -0.01)$
    (d) $P(Z < -0.79)$

3. For the standard normal distribution $Z \sim N(0, 1^2)$, find the following to 4 decimal places. Use the fact that $P(a < Z < b) = P(Z < b) - P(Z < a)$
    (a) $P(1 < Z < 2)$
    (b) $P(0.56 < Z < 0.78)$
    (c) $P(-0.2 < Z < 0.3)$
    (d) $P(-1.96 < Z < -1.645)$

4. The random variable $X \sim N(0.75, 0.04^2)$. For each of the following values of x, write down the corresponding value of z from the standardised normal distribution $Z \sim N(0, 1^2)$.
    (a) $x = 0.75$
    (b) $x = 0.79$
    (c) $x = 0.68$
    (d) $x = 0.73$

Exercise 9E...

5. The normal distribution $X \sim N(100, 16^2)$. Write in terms of $\Phi(a)$ where a is a value to be found in each case:
   (a) $P(X < 120)$
   (b) $P(X < 92)$
   (c) $P(X > 108)$
   (d) $P(96 < X < 112)$
      Hint: write this in the form $\Phi(a) - \Phi(b)$

6. (a) (i) Use the normal probability table in your formula booklet to find the value of a such that $\Phi(a) = 0.8810$
      (ii) Find $P(Z > 1.18)$
   (b) In an exam, the top 11.9% of candidates are awarded the A* grade. Given that the mean score is 54.6% and the standard deviation is 30%, find the grade boundary for the A* grade.

7. (a) Use the normal probability table in your formula booklet to find the values of z that correspond to the interquartile range, 25% to 75%.
   (b) The total annual rainfall is recorded in a town in County Down over a long period of time. The amounts of rainfall in millimetres can be modelled using a normal distribution $R \sim N(890, 130^2)$. If the annual rainfall does not fall within the long-term interquartile range, it is considered an abnormal year. Find the range of rainfall amounts that would make for an:
      (i) abnormally dry year;
      (ii) abnormally wet year.

9.6 Finding The Mean And Standard Deviation

Given certain statistics about a normal distribution, you may be able to find either the mean μ, the standard deviation σ or both.

If both the mean and the standard deviation are required, you will usually have to solve simultaneous equations.

The following two examples use the normal probability table. See section 9.8 for example uses of the table.

In the following example, the unknown mean μ is found.

Worked Example

7. The random variable $X \sim N(\mu, 5^2)$
   Given that $P(X < 22) = 0.115$, find μ.

$$x = 22 \Rightarrow z = \frac{22 - \mu}{5}$$

$$P(X < 22) = 0.115 \Rightarrow P\left(Z < \frac{22 - \mu}{5}\right) = 0.115$$

We cannot use the normal probability table for probabilities less than 0.5.

Use the fact that $P(Z < -a) = 1 - P(Z < a)$

$$P\left(Z < \frac{\mu - 22}{5}\right) = 0.885$$

Using a "reverse lookup" in the table:

$$\frac{\mu - 22}{5} = 1.20$$

$$\mu = 28$$

Note: This can also be done using the **inverse normal** mode on the calculator.

In the following example, the unknown standard deviation σ is found.

Worked Example

8. A machine makes metal rods to be used as tool components. The lengths of these rods, measured in millimetres, can be modelled as a random variable X that follows a normal distribution, $X \sim N(60, \sigma^2)$. If a rod is chosen at random, it is known that $P(X < 56.5) = 0.2420$ to 4 decimal places.
   (a) Find σ.
   (b) Find the upper quartile for the rod lengths.

(a) $P(X < 56.5) = 0.2420$

   The corresponding z value is:

   $$z = \frac{56.5 - 60}{\sigma} = -\frac{3.5}{\sigma}$$

   Therefore:

   $$P\left(Z < -\frac{3.5}{\sigma}\right) = 0.2420$$

   Since this is a probability less than 0.5, use the fact that $P(Z < -a) = 1 - P(Z < a)$

   So:

   $$P\left(Z < \frac{3.5}{\sigma}\right) = 0.7580$$

Using a reverse lookup, $\dfrac{3.5}{\sigma} = 0.7$

$$\Rightarrow \sigma = \dfrac{3.5}{0.7} = 5$$

(b) Three-quarters of the rod lengths are smaller than the upper quartile. So, $P(X < x) = 0.75$ where x is the upper quartile.

Find a corresponding z value:

$P(Z < z) = 0.75 \Rightarrow z = 0.674$

> **Note:** See Worked Example 13(d) later in this chapter for instructions on finding a z value when the probability lies between two values given in the table.

$$z = \dfrac{x - \mu}{\sigma}$$

$$\Rightarrow 0.674 = \dfrac{x - 60}{5}$$

$$\Rightarrow x = 63.37$$

The upper quartile Q_3 is 63.4 mm (3 s.f.).

....

In the next example, two pieces of information about the distribution have been given, allowing simultaneous equations to be formed to find both the mean and the standard deviation.

....

Worked Example

9. The masses of the women in a town are normally-distributed with a mean of μ kg and a standard deviation of σ kg. The probability that a woman chosen at random has a mass of less than 74.16 kg is 0.8106. The probability that a woman chosen at random has a mass of more than 53.16 kg is 0.9830. Find the mean μ and the standard deviation σ.

Write down the information provided in the question:

$$P(X < 74.16) = 0.8106 \qquad (1)$$
$$P(X > 53.16) = 0.9830 \qquad (2)$$

From (1):

$$P(X < 74.16) = 0.8106$$

$$\Rightarrow P\left(Z < \dfrac{74.16 - \mu}{\sigma}\right) = 0.8106$$

$$\Rightarrow \Phi\left(\dfrac{74.16 - \mu}{\sigma}\right) = 0.8106$$

Using a reverse look-up in the normal distribution table:

$$\dfrac{74.16 - \mu}{\sigma} = 0.88$$

$$74.16 - \mu = 0.88\sigma \qquad (3)$$

From (2):

$$P(X > 53.16) = 0.9830$$

$$P(X < 53.16) = 0.0170$$

$$P\left(Z < \dfrac{53.16 - \mu}{\sigma}\right) = 0.0170$$

$$\Rightarrow \Phi\left(\dfrac{53.16 - \mu}{\sigma}\right) = 0.0170$$

Use the fact that $\Phi(z) = 1 - \Phi(-z)$. By getting the right hand side > 0.5 we can perform the reverse look-up in the normal probability table:

$$\Rightarrow \Phi\left(\dfrac{\mu - 53.16}{\sigma}\right) = 0.9830$$

$$\dfrac{\mu - 53.16}{\sigma} = 2.12$$

$$\mu - 53.16 = 2.12\sigma \qquad (4)$$

Solve (3) and (4) as simultaneous equations:

$$\Rightarrow 21 = 3\sigma$$
$$\sigma = 7$$

Substitute into (4):

$$\Rightarrow \mu = 68$$

....

Exercise 9F

1. The random variable $X \sim N(\mu, 4^2)$ and $P(X < 15) = 0.9032$. Find the mean μ.

2. The random variable $Y \sim N(26, \sigma^2)$ and $P(Y < 29) = 0.7734$. Find the standard deviation σ.

3. The weight of a certain breed of small dog in newtons can be modelled using the random variable $W \sim N(\mu, 10^2)$. Given that $P(W > 130) = 0.1611$. Find the mean μ.

4. The random variable $Q \sim N(48, \sigma^2)$ and $P(Q > 56) = 0.2611$. Find the standard deviation σ.

5. The random variable $X \sim N(\mu, 12^2)$ and $P(X < 20) = 0.0694$. Find the mean μ.

6. The random variable $Y \sim N(3000, \sigma^2)$ and $P(Y < 2750) = 0.1469$. Find the standard deviation σ.

Exercise 9F...

7. It is known that 10% of bags of potatoes weigh less than 490 grams and 20% weigh more than 505 grams. If the masses of these bags is known to be normally-distributed, find, to 4 significant figures, the mean μ and standard deviation σ in grams.

8. A 5 km race is held in Dungannon. The times taken by the runners are normally-distributed with a mean of μ minutes and a standard deviation of σ minutes. The probability that a runner chosen at random has a time of less than 30 minutes is 0.9911. The probability that a runner chosen at random has a time of more than 20 minutes is 0.9484. Find the mean μ and the standard deviation σ.

9. A company in Carrickfergus employs travelling salespeople. The distances these salespeople drive around Northern Ireland can be modelled as a normal distribution $D \sim N(\mu, \sigma^2)$. Given that 10% of the journeys are 18 km or less and that 5% of the journeys are 30 km or more,
   (a) Sketch a diagram to represent this information.
   (b) Find the value of μ and the value of σ.
   (c) The company manager chooses one of the journeys at random. Find the probability it was a journey of more than 25 km.

10. During the summer months of June, July and August, Gráinne's solar panels generate a mean of μ kW h of electricity each day, with a standard deviation of σ kW h. Gráinne notices that her solar panels generate less than 16 kW h of electricity on 20% of summer days. She also observes that on 10% of summer days, the electricity generated is more than 18 kW h. Assuming that the daily amount can be modelled as a random variable, distributed normally, find:
    (a) The value of μ and the value of σ.
    (b) The interquartile range.

9.7 Calculator Methodologies

The tips here are written for the Casio fx-991EX *Classwiz*, but the steps are similar on other Casio calculators that have this functionality.

Normal cumulative distribution
You can find probabilities relating to the normal distribution on your calculator. To find $P(x < X < y)$ on your calculator, follow these steps:

1. Press **MENU**, then **7** to enter Distribution mode.
2. Press **2** to enter Normal Cumulative Distribution.
3. Enter the lower bound, x, then press =.
4. Enter the upper bound, y, then press =.
5. Enter the standard deviation, σ, then press =.
6. Enter the mean, μ, then press =.
7. Press = again. The calculator display shows the value of P.

Worked Example
10. Given $X \sim N(20, 5^2)$, find:
    (a) $P(18 < X < 21)$
    (b) $P(X < 23)$
    (c) $P(X > 25)$

    (a) To find $P(18 < X < 21)$, follow the steps above with a lower bound of 18, an upper bound of 21, $\sigma = 5$ and $\mu = 20$.
    The calculator gives $P = 0.2346814511$
    So $P(18 < X < 21) = 0.235$ (to 3 s.f.)

    (b) To find $P(X < 23)$, the steps involved are similar to those above, but in this case a lower bound has not been given.
    The probability required is the area beneath the curve between $-\infty$ and 23, i.e. $P(-\infty < X < 23)$. In place of $-\infty$, use a very small number for the lower bound, for example -1000. Since the area in the left- and right-hand tails of the distribution is negligible, this will provide a very good approximation of the true area.
    The calculator gives $P = 0.7257468853$
    So $P(X < 23) = 0.726$ (3 s.f.)

    (c) To find $P(X > 25)$, the steps are similar again, this time using a lower bound of 25 and an upper bound of 1000, or any large number.
    The calculator gives $P = 0.1586552539$
    So $P(X > 25) = 0.159$ (3 s.f.)

Inverse normal distribution

Given a probability p you can use the Inverse Normal function on your calculator to find a, such that
$P(X < a) = p$

On the Casio fx-991EX calculator, the probability is referred to as an area, since it represents an area under the curve to the left of a.

Given a probability, find a on your calculator by following these steps:

1. Press **MENU**, then **7** to enter Distribution mode.
2. Press **3** to enter Inverse Normal.
3. Enter the area (the probability p), then press =.
4. Enter the standard deviation, σ, then press =.
5. Enter the mean μ, then press =.
6. Press = again. The calculator display shows the value xInv (x).

Worked Example

11. Given $X \sim N(100, 5^2)$, find the value of a such that:
    (a) $P(X < a) = 0.75$
    (b) $P(X > a) = 0.4$
    (c) $P(90 < X < a) = 0.3$

(a) Since the area under the curve and to the left of a is 0.75, the following sketch shows that the value of a is greater than 100.

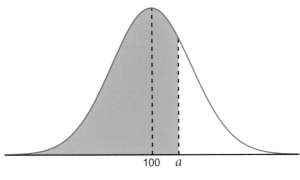

To find a where $P(X < a) = 0.75$, follow the steps above with area = 0.75, $\sigma = 5$ and $\mu = 100$.

The calculator gives xInv = 103.3724479
So $a = 103$ (3 s.f.)

(b) The area $P(X > a) = 0.4$ is shown in the following sketch.

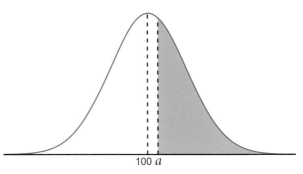

$P(X > a) = 0.4$ so $P(X < a) = 0.6$
Use the steps above, this time with an area of 0.6.

The calculator gives xInv = 101.2667355
So $a = 101$ (3 s.f.)

(c) The area $P(90 < X < a) = 0.3$ is shown in the following sketch.

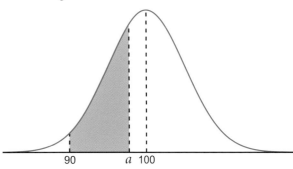

Use the fact that
$P(90 < X < a) = P(X < a) - P(X < 90)$
Then:
$$P(X < a) - P(X < 90) = 0.3$$
$$P(X < a) = P(X < 90) + 0.3$$

Using the normal cumulative distribution function on the calculator gives:
$$P(X < 90) = 0.02275$$
$$P(X < a) = 0.02275 + 0.3$$
$$= 0.32275$$

Using the inverse normal mode again gives:
xInv = 97.69988725
So $a = 97.7$ (3 s.f.)

9.8 How To Use The Normal Probability Table

As an alternative to using the calculator, you can use the normal probability table in your formula booklet.

The table gives $\Phi(z)$, or $P(Z < z)$ for values of z from 0 to 3.99.

The four parts of the next example show how to use the table as an alternative to using the **normal cumulative distribution** function on the calculator.

Worked Example

13. $X \sim N(20, 5^2)$

Find to 4 decimal places:

(a) $P(X < 22.1)$

(b) $P(X < 15.8)$

(c) $P(X > 20.7)$

(d) $P(X < 22.46)$

(a) Calculate the z value corresponding to $x = 22.1$:

$$z = \frac{x - \mu}{\sigma} = \frac{22.1 - 20}{5} = 0.42$$

Look up the probability associated with a z value of 0.42:

z	0.00	0.01	0.02	0.03	0.04	0.05
0.0	0.5000	0.5040	0.5080	0.5120	0.5160	0.5199
0.1	0.5398	0.5438	0.5478	0.5517	0.5557	0.5596
0.2	0.5793	0.5832	0.5871	0.5910	0.5948	0.5987
0.3	0.6179	0.6217	0.6255	0.6293	0.6331	0.6368
0.4	0.6554	0.6591	0.6628	0.6664	0.6700	0.6736
0.5	0.6915	0.6950	0.6985	0.7019	0.7054	0.7088

$P(X < 22.1) = 0.6628$

(b) Calculate the z value corresponding to $x = 15.8$:

$$z = \frac{x - \mu}{\sigma} = \frac{15.8 - 20}{5} = -0.84$$

We cannot look up the probability associated with this z value, since the table includes z values greater than or equal to 0. Instead, use $z = 0.84$.

z	0.00	0.01	0.02	0.03	0.04	0.05
0.0	0.5000	0.5040	0.5080	0.5120	0.5160	0.5199
0.1	0.5398	0.5438	0.5478	0.5517	0.5557	0.5596
0.2	0.5793	0.5832	0.5871	0.5910	0.5948	0.5987
0.3	0.6179	0.6217	0.6255	0.6293	0.6331	0.6368
0.4	0.6554	0.6591	0.6628	0.6664	0.6700	0.6736
0.5	0.6915	0.6950	0.6985	0.7019	0.7054	0.7088
0.6	0.7257	0.7291	0.7324	0.7357	0.7389	0.7422
0.7	0.7580	0.7611	0.7642	0.7673	0.7704	0.7734
0.8	0.7881	0.7910	0.7939	0.7967	0.7995	0.8023
0.9	0.8159	0.8186	0.8212	0.8238	0.8264	0.8289

The table tells us that $P(Z < 0.84) = 0.7995$.

Then use the fact that:

$$P(Z < -a) = 1 - P(Z < a)$$
$$P(Z < -0.84) = 1 - P(Z < 0.84)$$
$$= 1 - 0.7995$$
$$= 0.2005$$

(c) Calculate the z value corresponding to $x = 20.7$.

$$z = \frac{x - \mu}{\sigma} = \frac{20.7 - 20}{5} = 0.14$$

Look up the probability:

z	0.00	0.01	0.02	0.03	0.04	0.05
0.0	0.5000	0.5040	0.5080	0.5120	0.5160	0.5199
0.1	0.5398	0.5438	0.5478	0.5517	0.5557	0.5596
0.2	0.5793	0.5832	0.5871	0.5910	0.5948	0.5987
0.3	0.6179	0.6217	0.6255	0.6293	0.6331	0.6368
0.4	0.6554	0.6591	0.6628	0.6664	0.6700	0.6736
0.5	0.6915	0.6950	0.6985	0.7019	0.7054	0.7088

The table tells us that $P(Z < 0.14) = 0.5557$.
Use the fact that:

$$P(Z > a) = 1 - P(Z < a)$$
$$P(Z > 0.14) = 1 - 0.5557 = 0.4443$$

So:

$$P(X > 20.7) = P(Z > 0.14) = 0.4443$$

(d) Calculate the z value corresponding to $x = 22.46$.

$$z = \frac{x - \mu}{\sigma} = \frac{22.46 - 20}{5} = 0.492$$

Since the z value has 3 decimal places, look in the ADD section of the table.

Look up the probability for $z = 0.49$ and add 7 to the final decimal place.

0.07	0.08	0.09	(ADD) 1	2	3	4	5	6	7	8	9
0.5279	0.5319	0.5359	4	8	12	16	20	24	28	32	36
0.5675	0.5714	0.5753	4	8	12	16	20	24	28	32	36
0.6064	0.6103	0.6141	4	8	12	15	19	23	27	31	35
0.6443	0.6480	0.6517	4	8	11	15	19	23	26	30	34
0.6808	0.6844	0.6879	4	7	11	14	18	22	25	29	32
0.7157	0.7190	0.7224	3	7	10	14	17	21	24	27	31

This tells us that:

$$P(Z < 0.492) = 0.6879 + 0.0007 = 0.6886$$

So:

$$P(X < 22.46) = P(Z < 0.492) = 0.6886$$

The four parts of the following example show how to use the table as an alternative to using the **inverse normal** function on the calculator.

Worked Example

14. $X \sim N(20, 5^2)$

Find x, given that:

(a) $P(X < x) = 0.6026$
(b) $P(X < x) = 0.3669$
(c) $P(X > x) = 0.3015$
(d) $P(X < x) = 0.6999$

In these questions we have been given a probability (or area). We use the table for a 'reverse lookup', that is we read off the z value for the corresponding probability. Then we convert this z value to an x value using $z = \dfrac{x - \mu}{\sigma}$

(a) For $P(X < x) = 0.6026$, first look up the corresponding z value in the table:

z	0.00		0.03	0.04	0.05	0.06
0.0	0.5000		0.5120	0.5160	0.5199	0.5239
0.1	0.5398		0.5517	0.5557	0.5596	0.5636
0.2	0.5793		0.5910	0.5948	0.5987	0.6026
0.3	0.6179		0.6293	0.6331	0.6368	0.6406
0.4	0.6554		0.6664	0.6700	0.6736	0.6772
0.5	0.6915		0.7019	0.7054	0.7088	0.7123

The table tells us that, for a probability of 0.6026, $z = 0.26$

Then using:

$$z = \frac{x - \mu}{\sigma}$$

$$0.26 = \frac{x - 20}{5}$$

$$x = 5(0.26) + 20$$

$$= 21.3$$

(b) For $P(X < x) = 0.3669$, the probability is less than 0.5, but the table only includes probabilities of 0.5 and greater. Probabilities less than 0.5 correspond to negative z values.

Use the fact that:

$P(Z < -a) = 1 - P(Z < a)$

So:

$P(Z < -z) = 0.6331$

Look up this probability in the table:

z	0.00	0.01	0.02	0.03	0.04	0.05
0.0	0.5000	0.5040	0.5080	0.5120	0.5160	0.5199
0.1	0.5398	0.5438	0.5478	0.5517	0.5557	0.5596
0.2	0.5793	0.5832	0.5871	0.5910	0.5948	0.5987
0.3	0.6179	0.6217	0.6255	0.6293	0.6331	0.6368
0.4	0.6554	0.6591	0.6628	0.6664	0.6700	0.6736
0.5	0.6915	0.6950	0.6985	0.7019	0.7054	0.7088

From the table, $-z = 0.34$, giving $z = -0.34$

Then using:

$$z = \frac{x - \mu}{\sigma}$$

$$-0.34 = \frac{x - 20}{5}$$

$$x = 5(-0.34) + 20$$

$$= 18.3$$

(c) For $P(X > x) = 0.3015$ use the fact that:

$P(Z < a) = 1 - P(Z > a)$

$P(Z < z) = 0.6985$

Look up this probability in the table:

z	0.00	0.01	0.02	0.03	0.04	0.05
0.0	0.5000	0.5040	0.5080	0.5120	0.5160	0.5199
0.1	0.5398	0.5438	0.5478	0.5517	0.5557	0.5596
0.2	0.5793	0.5832	0.5871	0.5910	0.5948	0.5987
0.3	0.6179	0.6217	0.6255	0.6293	0.6331	0.6368
0.4	0.6554	0.6591	0.6628	0.6664	0.6700	0.6736
0.5	0.6915	0.6950	0.6985	0.7019	0.7054	0.7088
0.6	0.7257	0.7291	0.7324	0.7357	0.7389	0.7422

The table gives $z = 0.52$

Then using:

$$z = \frac{x - \mu}{\sigma}$$

$$0.52 = \frac{x - 20}{5}$$

$$x = 22.6$$

(d) For $P(X < x) = 0.6999$, this probability lies between two probabilities in the table:

z	0.00	0.01	0.02	0.03	0.04	0.05
0.0	0.5000	0.5040	0.5080	0.5120	0.5160	0.5199
0.1	0.5398	0.5438	0.5478	0.5517	0.5557	0.5596
0.2	0.5793	0.5832	0.5871	0.5910	0.5948	0.5987
0.3	0.6179	0.6217	0.6255	0.6293	0.6331	0.6368
0.4	0.6554	0.6591	0.6628	0.6664	0.6700	0.6736
0.5	0.6915	0.6950	0.6985	0.7019	0.7054	0.7088
0.6	0.7257	0.7291	0.7324	0.7357	0.7389	0.7422

A probability of 0.6985 corresponds to a z value of 0.52.

Look at the ADD section for this row of the table:

(ADD)								
1	2	3	4	5	6	7	8	9
4	8	12	16	20	24	28	32	36
4	8	12	16	20	24	28	32	36
4	8	12	15	19	23	27	31	35
4	8	11	15	19	23	26	30	34
4	7	11	14	18	22	25	29	32
3	7	10	14	17	21	24	27	31

Since the required probability of 0.6999 is 0.6985 + 0.0014, read off 4 from the top row of the ADD table. This is the third decimal place of the z value.

$z = 0.524$

Then using:

$$z = \frac{x - \mu}{\sigma}$$

$$0.524 = \frac{x - 20}{5}$$

$$x = 22.62$$

9.9 Summary

The area under a continuous probability distribution curve is equal to 1.

If X is a normally-distributed random variable you write $X \sim N(\mu, \sigma^2)$, where μ is the population mean and σ is the population standard deviation.

The normal distribution curve:

- Has parameters μ, the population mean, and σ^2 the population variance.
- Is symmetrical.
- Has a bell-shaped curve with an asymptote at each end.
- Has a total area under the curve equal to 1

The **standard normal distribution** has mean zero and standard deviation 1. The standard normal variable is written as $Z \sim N(0, 1^2)$

You can find probabilities for the normal distribution on your calculator or using the Normal Probability table in the formula booklet.

Given a probability you can use the table to find a value of z. You can also use **inverse normal** mode on your calculator to find a value of x or z, given a probability.

You can solve equations to find μ and σ given information about the distribution.

Chapter 10
Working With Probability

10.1 Introduction

In A2 Applied Mathematics you will be asked to select an appropriate distribution for a given scenario.

In this chapter you will learn when and how to use the normal distribution or whether another distribution, such as the binomial distribution, may be more appropriate.

You should understand when the binomial and the normal distributions are not applicable.

You need to recognise the context for a binomial or a normal distribution – for example, is the data discrete or continuous etc? You also need to understand the conditions required for each of these distributions.

This chapter also discusses further modelling using probability distributions.

> **Note:** The student does not need to know the normal approximation to the discrete binomial distribution in CCEA A2 Applied Mathematics. As such, knowledge of the continuity correction is not required.

Key words
- **Normal distribution**: A probability distribution characterized by a bell-shaped curve. The normal distribution occurs a lot in natural processes.
- **Binomial distribution**: A probability distribution, in which the statistic is the number of 'successes' in a fixed number of trials.
- **Discrete**: A discrete random variable can only take certain values, for example the outcome of rolling a single die. The binomial distribution is an example of a discrete distribution.
- **Continuous**: A continuous random variable can take any value within a range, for example the height of a randomly chosen person. The normal distribution is an example of a continuous random variable.

Before you start
You should:
- Know how to calculate probabilities for combinations of events, for example using a tree diagram.

- Understand that the normal distribution is an example of a continuous probability distribution.
- Understand that the binomial distribution is an example of a discrete probability distribution.
- Understand that a model is a representation of a real-life process and that a model can never give results that exactly replicate measurements seen in real-life.

Key characteristics of the normal and binomial distributions
By way of revision, the following are the key characteristics of the normal and binomial distributions.

The normal distribution:
- Has parameters μ, the population mean, and σ^2, the population variance, so we write $X \sim N(\mu, \sigma^2)$.
- Is a symmetrical distribution, meaning that the mean and the median are equal.
- Has a bell-shaped curve with asymptotes at both ends.
- Has a total area under the curve of 1.

The shape of the distribution is shown in the diagram below.

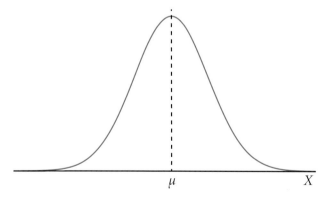

The mean is the value of X that corresponds to the centre of the curve. It is the value that is associated with the greatest probability.

Notes about the curve:
- The important aspects of the graph are the numbers on the horizontal axis and the area below the curve, which represents probability.
- Although a normal random variable could take any value, in practice it very rarely lies more than

3 standard deviations from the mean, so for values of X at these distances from the mean, the probability density falls very close to zero.

- Approximately 68% of the data lies within one standard deviation of the mean.
- 95% of the data lies within two standard deviations of the mean.
- Nearly all the data (99.7%) lies within three standard deviations of the mean.

Examples of random variables that are distributed normally:

- The heights of people within Northern Ireland.
- The masses of chimpanzees in a forest.
- The level of blood glucose in a certain population.
- The masses of breakfast cereal in boxes that are labelled '750 grams'.

....

Worked Example

1. The heights of people within Northern Ireland may be modelled using a normal distribution, with a mean of 165 cm and a standard deviation of 11 cm. If X is the random variable 'the height of a randomly chosen person in Northern Ireland':
   (a) Write down the probability distribution for X;
   (b) Sketch the probability distribution.

   (a) In general, the normal probability distribution is
   $X \sim N(\mu, \sigma^2)$

   In this case we can write:
   $X \sim N(165, 11^2)$

   (b) The distribution can be shown on a graph as below.

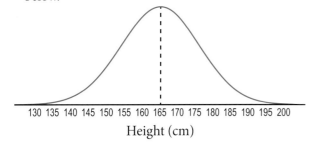

   Height (cm)

....

The binomial distribution

The binomial distribution has parameters n, the number of trials, and p, the probability of success in each trial. So, we write $X \sim B(n, p)$.

Criteria for a random variable X to be modelled using a binomial distribution are:

- A fixed number of trials.
- Independent trials.
- There should be two mutually exclusive and exhaustive outcomes for each trial (sometimes called 'success' and 'failure').

> **Note:** Mutually exclusive means they can't both happen. Exhaustive means one of them must happen.

- A fixed probability of success for each trial.
- X is the number of 'successes'.

The fact that there is a fixed number of trials means that there is always an upper limit to the value that X can take.

Examples of random variables that are distributed binomially:

- The number of heads that come up when a coin is tossed 20 times.
- The number of seeds that germinate out of 30 seeds planted.
- The number of patients who have diabetes out of the 50 patients treated by a doctor in one day.

Exercise 10A (Revision)

1. A fair coin is tossed 3 times. By drawing a tree diagram or otherwise, find the probability of getting two heads and a tail.

2. Write down two key characteristics of the normal distribution curve.

3. The binomial distribution is an example of a discrete probability distribution. Write down three requirements for a discrete random variable to be modelled using the binomial distribution.

4. A zoologist studies the wingspans of a certain species of bat. The distribution of these wingspans can be modelled using a normal distribution with a mean of 80 mm and a standard deviation of 6 mm.
   (a) Write down the probability distribution.
   (b) Sketch the distribution.

5. A model for a tree's height h, in metres, after t years can be given by: $h^2 = 2.5t + 2$.
   A tree has a height of 5.1 metres after 10 years. Do you think the model is suitable? Justify your answer.

What you will learn

In this chapter you will learn how to:

- Model with probability, including critiquing assumptions made and the likely effect of more realistic assumptions.
- Select an appropriate probability distribution for a context, with appropriate reasoning, including recognising when a binomial or normal model may not be appropriate.

In the real world...

The Met Office forecasts the weather using a sophisticated computer model. The model uses input data from buoys at sea, satellites, ground-based weather stations and other sources. However, the output from a weather forecasting model is highly dependent on the data that is put into the model.

There is always some uncertainty in these input data. Were the measurements taken accurately? Is there missing data, such as might occur when equipment breaks down? Even the level of rounding for a reading can make a difference.

So, for example, changing a sea surface temperature off the coast of Ireland, which is used as an input to the model, can affect the amount of rainfall predicted during the next few days.

One way in which the Met Office has improved its forecasts is by using ensemble forecasts. Using this approach, the model is run several times, each time using a slightly different set of input data, to account for the fact that there is uncertainty in these inputs.

The ensemble forecasts give the forecaster a much better idea of what weather events may occur at a particular time. By comparing the different forecasts in the ensemble, the forecaster can decide how likely a particular weather event will be. If the forecasts vary a lot, then the forecaster knows that there is a lot of uncertainty about what the weather will do, but if the forecasts are all very similar, they will have more confidence in predicting a particular event.

The job of the TV or radio weather presenter is to convey the central, most likely scenario, but also to express the level of certainty that they have in the forecast, based on the spread of the ensemble results.

10.2 Selecting An Appropriate Probability Distribution For A Context

Selecting an appropriate distribution

In A2 Applied Mathematics, you are required to recognise when a binomial or normal model may or may not be appropriate, giving appropriate reasoning.

Worked Examples

2. A rattlesnake lays 20 eggs. X is the random variable 'the number of eggs that hatch'.
   (a) Give one reason X **cannot** be modelled as a normal random variable.
   (b) Explain why X **can** be modelled as a binomial random variable.
   (c) Give one reason that the binomial model may be slightly inaccurate.

   (a) X is not normal since it is a discrete random variable.

   (b) X can best be modelled as a binomial random variable, since:
   - there is a fixed number of trials, 20;
   - the trials are independent;
   - there are two possible outcomes for each trial ('hatches' and 'doesn't hatch');
   - there is a fixed probability of success for each trial, perhaps a probability of 0.5 that an egg hatches, although see the caveat in part (c) below.

   (c) It may not be completely accurate that there is a fixed probability of an egg hatching. For example, some eggs may be smaller or less viable than others. There would then be a lower probability of 'success' for these trials.

3. A die is rolled until a 6 appears. X is a random variable whose value is the number of rolls of the die.
   (a) Draw a tree diagram showing how to calculate $P(X = 1), P(X = 2)$ and $P(X = 3)$.
   (b) Does X follow a normal distribution, a binomial distribution, or neither? Give a brief reason for your answer.

(a)

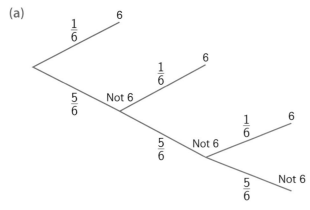

$$P(X = 1) = \frac{1}{6}$$

$$P(X = 2) = \frac{5}{6} \times \frac{1}{6} = \frac{5}{36}$$

$$P(X = 3) = \frac{5}{6} \times \frac{5}{6} \times \frac{1}{6} = \frac{25}{216}$$

(b) X is neither normally distributed, nor binomially distributed.

X is a discrete random variable, so it cannot be normally distributed.

It cannot be modelled using the binomial distribution. One valid explanation is that there is no upper limit to the value of X – in theory the die could be rolled indefinitely without getting a 6.

Other explanations are also available. For example, X does not arise from a binomial trial (one with two outcomes corresponding to success and failure).

> **Note:** Where successive probabilities are multiplied by a constant, as in this case where the multiplying factor is ⅚, this is known as a geometric distribution. A detailed study of this distribution was not a part of the CCEA specification at the time of publication.

Exercise 10B

1. (a) Give a reason why the normal distribution is important in statistics.
   (b) Give an example of a random variable that can be modelled using a normal distribution.

Exercise 10B...

2. State, with a reason, whether these random variables are discrete or continuous.
   (a) X, the length of a random sample of beetles.
   (b) Y, the scores achieved by 100 students in a maths test.
   (c) M, the masses of each cow in a herd of 200.
   (d) S, the collar sizes of the 1500 shirts sold in a clothes shop during a month.

3. In each case below, state whether X could be a random variable modelled using a normal probability distribution, a binomial distribution, or neither.
   (a) Rolling one die 10 times. X is the number of sixes.
   (b) X is the diameter of circular machine components being produced in a factory. Each component is supposed to have a diameter of 22 mm.
   (c) Tossing a coin 40 times. X is the number of heads.
   (d) X is the weights of newborn babies in Antrim Area Hospital during the month of August.
   (e) X is the time taken by a runner selected at random in the Belfast Marathon.
   (f) A student's name is drawn at random from a class of 25. X is the name.

4. There are ten beads in a bag, three white and seven black. For each part below, state whether the random variable X can be modelled using a binomial distribution, a normal distribution, or neither. Briefly justify your choice in each case.
   (a) A bead is chosen at random and replaced. This is repeated until four beads have been selected. X is the number of white beads chosen.
   (b) A bead is chosen at random, but not replaced. This is repeated until four beads have been selected. X is the number of white beads chosen.

5. A single, fair, six-sided die is rolled. X is the number shown on the die. State, giving a reason, whether X could be a random variable modelled using a normal probability distribution, a binomial distribution, or neither.

Exercise 10B...

6. A drug to treat asthma is claimed to be effective for 75% of sufferers. A doctor suspects that the effectiveness of the drug has been overstated. She takes a random sample of 25 of her patients whom she has treated with the drug and she finds that the drug was effective on 17 of them. What assumptions need to be made for a binomial model to be appropriate in this context?

7. Twenty chickens are in a hen coop.
   (a) State, giving a reason, whether the random variables described below could be modelled using a normal probability distribution, a binomial distribution, or neither.
      (i) X is the number of hens that lay eggs in a day.
      (ii) Y is the number of eggs the hens lay between them in a day.
   (b) If you answered binomial or normal to either part (a)(i) or (a)(ii), state any modelling assumptions you have made.

8. The distribution of the salary in pounds per year of the employees in an office building is shown. Is the normal distribution a suitable model for this dataset? Justify your answer.

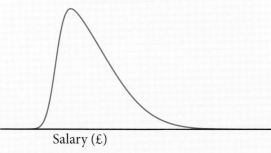

Salary (£)

9. In a forest, the heights of the trees H m can be modelled using a normal distribution, such that $H \sim N(4, 0.75^2)$
   (a) Explain why the median value $Q_2 = 4$ m
   (b) It is common practice in forest management to 'thin' the forest out by removing some of the smaller saplings to give the remaining saplings more space to mature.
      (i) How would thinning affect the mean?
      (ii) After thinning, do you think a normal distribution would still be an appropriate model for the distribution of the heights?

Exercise 10B...

10. Five fair, six-sided dice are rolled together. The random variable X is the **mean** score on the five dice. The diagram shows the probability distribution for X.

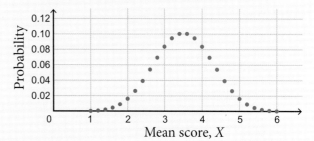

State, giving a reason, whether the random variable X is:
   (a) a normally distributed random variable;
   (b) a binomially distributed random variable; or
   (c) neither.

10.3 Modelling With Probability

In A2 Applied Mathematics you are required to model real-life processes with probability.

You need to make assumptions in forming a model, or state assumptions that have been made in a model that has been given.

You may be asked to critique an assumption, i.e. state whether an assumption is realistic. You may be asked to provide a more realistic assumption and state the effects this is likely to have on your model.

....

Worked Examples

4. Jake plans to toss a coin 30 times. He tells his sister Ella that he will get 15 heads and 15 tails.
   (a) State one modelling assumption Jake is making.
   (b) Comment further on Jake's prediction.

   (a) Jake has made the assumption that his coin is fair, i.e. that $P(\text{heads}) = P(\text{tails}) = 0.5$

   (b) There is always some uncertainty in the prediction of a model. So, it is not possible for Jake to be certain about the number of heads and tails, even if his modelling assumption that the coin is fair is correct.

5. Two 6-sided dice are thrown, one black and one white. X is the random variable 'the difference between the two numbers shown'. X is the highest number minus the lowest, so the outcome is always positive or zero.
   (a) Draw a sample space diagram showing the possible outcomes for X.
   (b) Given that X is less than 3, find the probability that at least one die lands on a four.
   (c) State one modelling assumption used in your calculations.

(a) The sample space diagram is shown below. Each point on the grid contains the difference of the two numbers shown on the dice.

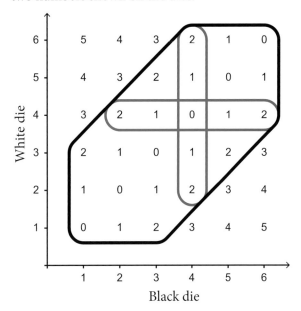

(b) The large black loop in the sample space diagram encloses all the ways in which a difference of less than 3 can be achieved. This is the restricted sample space. There are 24 outcomes in the restricted sample space. The nine grid points in the two blue loops represent one of the dice landing on a four.
   Therefore $P(\text{one } 4 \mid X < 3) = \dfrac{9}{24} = \dfrac{3}{8}$

(c) The modelling assumption is that both dice are fair.

Exercise 10C

1. Phil says that, when he tosses a coin, the chances of getting heads three times in a row is $\dfrac{1}{8}$. What modelling assumption is Phil making?

2. Siobhán plans to throw 12 six-sided dice. She says 'I will get two sixes'.
   (a) State one modelling assumption Siobhán is making.
   (b) Comment further on Siobhán's prediction.

3. Carol wants to have a baby girl. She continues to have children until she has a girl, then stops.
   (a) By drawing a probability tree diagram, or otherwise, find the probability she has exactly four children.
   (b) What modelling assumptions have been made in answering part (a)?

4. Deirdre chooses a book to read at random, with her eyes closed, from her bookshelf. She defines a random variable X to be the number of pages in the book chosen.

   Over 10 weeks, Deirdre repeats the process another 9 times. She finds $\bar{X}$, the mean of her ten X values.

   Deirdre notices that $\bar{X} > \mu$, where μ is the population mean. She does not know why her sample mean should be greater than the population mean and asks her brother Kevin, who is studying A2 Applied Mathematics.

   Kevin says that Deirdre was making an invalid modelling assumption. What was this modelling assumption?

5. Fred can be called into work any day of the week. If he is called, he should be at work at 9:00 a.m. He commutes into Belfast, leaving his house by car each morning. Fred knows from experience that the later he leaves, the more likely he is to be late. He models the probability of his being late using the formula:

   $$P(\text{late}) = 0.4 + \frac{T}{100}$$

   where T is the time he leaves his house, given as the number of minutes after 8:00 am.
   (a) According to Fred's model, what is the probability he will be late for work if he leaves his house at 8:15 am?
   (b) State one shortcoming with Fred's model.

Exercise 10C...

6. A spinner has five sections coloured red, orange, yellow, green and blue. Four of the sections each make an angle of 80° at the centre of the spinner, but the blue section makes an angle of 40°.

   (a) Aoife says that when she spins the spinner, the probability of it landing on blue is $\frac{1}{9}$. What modelling assumption is she making?

   (b) Assuming Aoife's modelling assumption is correct, find:

      (i) $P(\text{red})$

      (ii) $P(\text{red}|\text{not blue})$

7. Dan stops at his local coffee shop on his way to work in Strabane every morning. Each day he counts the number of people in the queue ahead of him. After 50 days he summarises his data: he observes that, on 24 days there was nobody in front of him; 18 days there was one person; 6 days there were two people; and on the remaining days there was a queue of more than two people.

   (a) Copy and complete the following table.

Length of queue in front of Dan (number of people)	Number of days	Relative frequency (estimate of probability)
0	24	$\frac{24}{50} = \frac{12}{25}$
1		
2		
>2		

   (b) If Dan goes to the coffee shop 200 times this year, how many times should he expect to find a queue of more than two people in front of him?

   (c) Dan estimates that if there is a queue of 2 or more people in front of him, there is a 50% chance he will be late for work. If there is a queue of less than 2, there is only a 10% chance he will be late. Find the probability Dan is late for work.

   (d) Briefly describe:

      (i) One modelling assumption Dan has used in his calculations.

      (ii) One way in which his modelling could be improved.

      (iii) How Dan could reduce the probability of being late for work.

8. Jarlath the football goalkeeper is involved in a penalty shoot-out. Five members of the opposing team take penalties against him. By looking at videos of previous shoot-outs, Jarlath has worked out the probabilities each player shoots to the left, right or centre and the probability of the player missing the goal altogether.

	Probability of player shooting:			Probability of player missing
	Left	Right	Centre	
Player 1	0.6	0.05	0.2	0.15
Player 2	0.55	0.2	0.15	0.1
Player 3	0.45	0.35	0.05	0.15
Player 4	0.1	0.7	0.1	0.1
Player 5	0.7	0.2	0.05	0.05

   For each penalty, Jarlath dives the way the penalty is most likely to go. If Jarlath dives the right way, there is a probability of 0.7 he will save the penalty.

   (a) Modelling the penalty shoot-out in this way, find the probability Jarlath saves the penalty from Player 1.

   (b) Hence show that the probability Player 1 **does not score** is 0.57.

   (c) For Jarlath's team to win the match, none of the opponents should score. Find the probability Jarlath's team wins using this model.

   (d) What modelling assumptions have been made?

9. Two six-sided dice, one black and one red, are thrown at the same time and the sum of the numbers shown is recorded as X.

   (a) Draw a sample space diagram for X.

   (b) Use your sample space diagram to help you find:

      (i) $P(X = 7)$

      (ii) $P(X = 10 \mid \text{Black die is 6})$

      (iii) $P(X = 3 \mid X \leq 5)$

   (c) What modelling assumptions have you made when answering this question?

Exercise 10C...

10. Rory has a six-sided die. X is the random variable 'the number shown on the die'.
    (a) Rory rolls his die three times. Let Y be the random variable 'the number of sixes rolled'. Rory is asked to find $P(Y = 3)$. He says that, when finding this probability, it is important to note that for his die $P(X = 6) = P(X < 6)$. What is Rory's answer?
    (b) What modelling assumption is Rory making?

11. The probability p of snow falling in a particular town on a winter's day is modelled using the formula:

    $$p^2 = \frac{5 - T}{100}$$

    where T is the air temperature in degrees Celsius (°C). Thirty years of data show that snow fell on Christmas Day three times and the average air temperature for these 30 Christmas Days was 4.1°C.
    (a) Based on this information, do you think the model is suitable? Justify your answer.
    (b) By considering the possible values of p that the model can produce, write down one shortcoming in the model.
    (c) In this model, the probability of snow falling depends on the air temperature alone. In a more sophisticated model, other variables may also appear on the right-hand side of the equation. Give an example of one of these variables.

10.4 Summary

In A2 Applied Mathematics you must be able to select an appropriate distribution for a given scenario. This includes being able to recognise when a binomial or normal model may or may not be appropriate, giving appropriate reasoning.

You need to recognise the context for a binomial or a normal distribution – for example, is the data discrete or continuous? You also need to understand the conditions required for each of these distributions to be a suitable model.

This chapter also discusses further modelling using probability.

You should be able to state any modelling assumptions that have been made in solving a problem using a probability model. You may also be asked to critique assumptions made and discuss the likely effect of more realistic assumptions.

Chapter 11
Statistical Hypothesis Testing

11.1 Introduction

This short chapter provides an introduction to the concept of statistical hypothesis testing. It introduces you to some of the important terminology and notation required.

In the following three chapters, we look in more depth at three specific types of hypothesis test:

- testing the mean in a normal distribution;
- testing the proportion in a binomial distribution; and
- testing the significance of a correlation coefficient.

Key words and notation

Throughout this chapter and the next three you will be introduced to some new terminology.

- **Parameter**: A quantity that relates to the underlying distribution, rather than being calculated from a sample. For example, population mean and population standard deviation are both parameters.
- **Statistic**: A quantity that does not relate to the underlying distribution; instead, it is usually calculated from a sample. For example, sample mean and sample standard deviation are both statistics.
- **Null Hypothesis**, H_0: What is currently considered to be true. The statement of H_0 is always an equality and always refers to a population parameter (as opposed to a sample statistic).
- **Alternative Hypothesis**, H_1: What might be true instead of H_0. This statement is **not** an equality, so it should involve an inequality sign, or a not equals sign: >, < or ≠
- **Population mean**, μ. The mean of all the possible values in the population. This is a theoretical value, i.e. a parameter. We usually do not know the true value of the population mean, as it is not possible to take the required measurement from every item in the population.
- **Sample mean**, $\bar{x}$: This is the mean of the values in the sample. This is a calculated value – a statistic.
- **Sample size**, n: The number of items in the sample.
- **Test statistic**: A value that is calculated from the sample. The test statistic is used as evidence in our decision whether to reject the null hypothesis.

Note: Different variables are used for the test statistic, depending on what type of test is being carried out.

- **Significance level**, α: The significance level is a measure of how confident we are that we reach the correct conclusion.
- **One-tailed test and two-tailed test**: A one-tailed test is used to determine whether a parameter has changed significantly in one direction, e.g. decreased. A two-tailed test is used to determine whether it has changed significantly in either direction (either increased or decreased).
- **Left-sided and right-sided**: A left-sided test is a one-tailed test in which we test whether a parameter is less than a particular value, for example a test for whether $\mu < 20$. A right-sided test is a one-tailed test in which we test whether a parameter is greater than a particular value, for example a test for whether $p > 0.8$.

Before you start

You should be familiar with:

- Basic probability theory and notation.
- The distinction between a population and a sample.

Exercise 11A (Revision)

1. A fair coin is tossed 3 times. Find the probability of getting:
   (a) 3 heads;
   (b) 1 tail and 2 heads.

2. In each of the following cases, explain why it is important to take a sample rather than a census.
   (a) Quality control in a sofa factory needs to be confident that the sofas are flame retardant.
   (b) A single researcher wishes to find out the annual incomes of the people in her town.
   (c) A researcher wishes to know the average length of a leaf from an oak tree.

What you will learn

In this chapter you will learn:

- The language of statistical hypothesis testing.
- That a sample is used to make an inference about the population.
- That the significance level is the probability of incorrectly rejecting the null hypothesis.

In the real world...

It is often said that exercise is good for your mental health. Is this true? And if so, does exercise have a short-term effect on your mental well-being, or is it something that accumulates over months and years? Is it simply because exercising helps to keep you healthy, and healthier people are happier?

Consider the following scenario: you go for a run in the park, and you feel great. Naturally, you might ask yourself 'does exercise make people happy?'. If you are asking a question that you don't know the answer to, research is necessary to resolve it. There are many forms that this research can take, from a literature review to performing an experiment. If an experiment is used, there are many different approaches.

To analyse the data and to answer this kind of question, researchers often turn to **statistical hypothesis testing**. It is a technique widely used in psychology, medicine, biological science and many other fields of research. It involves taking a sample and finding the probability of a test statistic. For example, 20 runners could be surveyed before and after a run and asked to assess their own level of happiness from one to ten. These data can be summarised as the test statistic.

But be careful! Using an incorrect or poorly-thought-out test can lead to wildly incorrect conclusions.

11.2 The Concept And Language Of Statistical Hypothesis Testing

The key words section at the beginning of this chapter introduces some of the important terminology that you will encounter in this chapter and the next three. This section expands on the meanings of some of these terms.

Population parameters

A **population parameter** is a quantity relating to the population in question. For example, in a population that follows a normal distribution, two population parameters are the mean (often denoted μ) and the standard deviation (often denoted σ).

In a population that follows a binomial distribution, a population parameter often discussed is p. p is the probability of success in a single trial, but it is also the **proportion** of trials that would result in success.

The null hypothesis and the alternative hypothesis

A **hypothesis** is a statement made about the value of a **population parameter**.

When performing a hypothesis test we form two opposing hypotheses: the **null** hypothesis and the **alternative** hypothesis. Both hypotheses relate to a **parameter**, for example the **population mean**.

The null hypothesis, denoted by H_0, is the default position: it states that the effect you are looking for does not exist. The alternative hypothesis, denoted by H_1, states that the suggested effect is taking place.

> The null hypothesis is denoted H_0. We assume this to be correct unless we find evidence to reject it.
>
> The alternative hypothesis is denoted H_1. We accept this if we reject H_0.

The sample and the test statistic

The goal of hypothesis testing is to collect evidence in the form of a **sample**.

The test statistic is the quantity that is under investigation and is derived from the sample. Examples include the number of accidents on a road, the number of sales of an item, etc. In most problems we are investigating whether some change has made a significant difference to the test statistic.

The sample is used, together with some assumptions about the distribution, to calculate the **test statistic**.

By looking at the value of the test statistic, we make a conclusion about the null hypothesis: we reject the null hypothesis if it appears unlikely to be true, i.e. if there is some experimental support for the alternative hypothesis.

It is important to note that, even if we accept the alternative hypothesis, we have not **proved** it to be true. We have gathered evidence to suggest that it is **likely** to be true.

....

Worked Example

1. A supermarket chain has 200 stores. In 2022, the team at Head Office develops a plan to improve sales by changing the layout of the stores. The following year, the team wishes to know whether the changes have had an effect. They sample 50 of the 200 stores.

(a) Assuming the distribution of the sales figures closely follows a normal distribution, what could be used as the test statistic?

Over several years, the average sales across all 200 stores has been £m.

(b) Suggest possible null and alternative hypotheses.

(a) When testing for a change in the mean μ in a normally distributed variable, the test statistic used is z_{test} which is calculated from the sample data.

> **Note:** Chapter 12 gives more information about calculating z_{test}.

(b) The following hypotheses would be suitable:

$H_0: \mu = m$

$H_1: \mu > m$

Alternatively, in words:

H_0: Mean sales at the stores have stayed the same.

H_1: Mean sales at the stores have increased.

The significance level

Assuming all the steps were carried out correctly, it is still possible to reach the wrong conclusion using a hypothesis test, because a hypothesis test is not a proof. Instead, we decide whether there is sufficient evidence to reject the null hypothesis, and therefore adopt the alternative hypothesis instead.

A fundamental part of any hypothesis test is the **significance level**. For example, you will see wording such as 'at the 5% significance level' or 'at the 10% significance level'. If we test at a 5% level of significance, there is a 5% chance that we reach the wrong conclusion. Put another way: there is a 95% chance we reach the right conclusion!

> The significance level is the probability of incorrectly rejecting the null hypothesis.

One-tailed and two-tailed tests

Chapter 12 will discuss testing the mean μ in a quantity that is normally distributed. In this case the null and alternative hypotheses will relate to μ, the population mean.

Chapter 13 will discuss testing the proportion p in a quantity that follows a binomial distribution. In this case the null and alternative hypotheses will relate to p, the proportion.

In either case, we can conduct either a one-tailed, or a two-tailed test.

Suppose we conduct a hypothesis test about a population parameter θ (where θ is either μ or p, depending on the type of test).

If we are testing whether θ has **increased**:

- The null hypothesis is $H_0: \theta = m$ (for some value m)
- The alternative hypothesis is $H_1: \theta > m$
- This is known as a **right-sided one-tailed test**.

If we are testing whether θ has **decreased**:

- The null hypothesis is the same as above, $H_0: \theta = m$
- The alternative hypothesis is $H_1: \theta < m$
- This is known as a **left-sided one-tailed test**.

In both cases, we conduct a **one-tailed test**, since we are investigating a change in θ in one direction only.

If we are testing whether θ has **changed**:

- The null hypothesis is the same as above, $H_0: \theta = m$
- The alternative hypothesis is $H_1: \theta \neq m$

This requires a **two-tailed test**, since we are investigating whether there has been a change in θ in either direction.

> **Note:** You may also come across the term **non-directional change**. A non-directional change could be in either direction, up or down, so if testing for non-directional change, use a two-tailed test.

Worked Examples

2. Patricia works in a lab testing water samples for bacteria. Over a long period of time, the mean concentration of bacteria in the River Lagan has been 950 per ml. There was a storm at the weekend, and much water drained into the river. Patricia suspects that the weekend storm has affected the concentration of bacteria in the water. To test Patricia's suspicion, 50 water samples are taken along the river. The mean concentration of bacteria in these samples is 1000 per ml. She carries out a hypothesis test at the 5% significance level.

   (a) State possible null and alternative hypotheses. State also whether this is a one-tailed or two-tailed test.

   (b) Assuming that Patricia carries out the test correctly, state the probability that she incorrectly rejects the null hypothesis.

   (a) Since Patricia suspects that the storm affected the concentration of bacteria, the correct hypotheses are:
   $H_0: \mu = 950; H_1: \mu \neq 950$

This is a two-tailed test, since the change in concentrations could be up or down.

However, a solution with the alternative hypothesis $H_1: \mu > 950$ may also be considered correct in an exam setting. This would then be a one-tailed test, since we are only testing for an increase.

(b) The probability of Patricia incorrectly rejecting the null hypothesis is 0.05, the significance level.

---

3. A coin has a probability p of coming up heads. Michael suspects the coin is biased towards heads. He tosses the coin 20 times and counts the number of times X that it comes up heads.
   (a) Describe the sample and the test statistic.
   (b) Should Michael conduct a one-tailed or two-tailed test?
   (c) Write down a suitable null hypothesis.
   (d) Write down a suitable alternative hypothesis.
   (e) If Michael carries out a test at a 5% significance level, what is the probability he comes to the correct conclusion about the coin?

(a) The sample is the 20 coin tosses and the test statistic that summarises his sample is the number of heads, X.
(b) Michael carries out a right-sided one-tailed test, as he is testing whether the probability p is greater than 0.5.
(c) If the coin is unbiased the probability of getting heads is 0.5 so the null hypothesis is: $H_0: p = 0.5$
(d) If the coin is biased towards heads, the probability of getting heads is greater than 0.5, so the alternative hypothesis is $H_1: p > 0.5$

> **Note:** If the question stated that Michael suspects the coin is biased towards tails, the alternative hypothesis becomes $H_1: p < 0.5$
> In this case, a one-tailed test is still required, but left-sided.
>
> If Michael suspects the coin is biased, but doesn't know which way, the alternative hypothesis becomes $H_1: p \neq 0.5$
> In this case, a two-tailed test is carried out.

(e) At the 5% significance level, there is a 95% chance Michael's conclusion is correct.

Exercise 11B

1. (a) Describe what is meant by a statistical hypothesis.
   (b) Describe the difference between the null hypothesis and the alternative hypothesis.
   (c) What symbols do we use to denote the null and alternative hypotheses?

2. (a) In the context of a statistical hypothesis test, explain clearly what is meant by the term 'one-tailed test'.
   (b) What are the two different types of one-tailed test?

3. In any statistical hypothesis test, there is a chance that the wrong conclusion may be reached. In a hypothesis test with a 10% significance level, James rejects the null hypothesis. What is the probability that the null hypothesis has been rejected incorrectly?

4. In a hypothesis test:
   (a) In what way is evidence gathered?
   (b) How is a test statistic calculated?

5. At a busy road intersection there were 32 traffic accidents during 2021. At the end of 2021 the road layout was changed, with a view to making the intersection safer. In 2022 the number of accidents falls to 28. What is the test statistic?

6. Farooq wants to see whether a die is biased towards the number six. He throws the die 60 times and counts the number of sixes he gets.
   (a) Describe the test statistic.
   (b) Should Farooq conduct a one-tailed or two-tailed test? If one-tailed, is it left-sided or right-sided?
   (c) Write down a suitable null hypothesis to test the die.
   (d) Write down a suitable alternative hypothesis.

7. Anya wants to test to see whether a coin is biased. She tosses the coin 100 times and counts the number of times she gets a head.
   (a) Describe the test statistic.
   (b) Should Anya conduct a one-tailed or two-tailed test? If one-tailed, is it left-sided or right-sided?
   (c) Write down a suitable null hypothesis to test the coin.
   (d) Write down a suitable alternative hypothesis.

Exercise 11B...

8. Over a long period of time, it is found that the mean number of accidents λ occurring at a crossroads is 5 per month. New traffic lights are installed. James works for the roads authority. His boss suggests that the rate of occurrence has changed in some way since this modification was made. He asks James to test his suggestion.

   (a) Describe the test statistic.

   (b) Should James conduct a one-tailed or two-tailed test? If one-tailed, is it left-sided or right-sided?

   (c) Write down a suitable null hypothesis James could use to test his boss's suggestion.

   (d) Write down a possible alternative hypothesis.

9. In a manufacturing process, the proportion p of faulty articles has been found, over a long period of time, to be 0.15. A new process is introduced, and the first batch of articles is sampled. The proportion of faulty articles in this batch is 0.12. The manufacturers wish to test, at the 5% significance level, whether or not there has been a reduction in the proportion of faulty articles.

   (a) Suggest a suitable test statistic.

   (b) Should a one-tailed or two-tailed test be carried out? If one-tailed, is it left-sided or right-sided?

   (c) Write down two suitable hypotheses.

   (d) Explain the condition under which the null hypothesis should be rejected.

10. On a turkey farm, the masses of the male turkeys are thought to be normally distributed. Over the long term the mean mass of a male turkey has been measured at 7.8 kg. In 2022, avian 'flu affected the population. The farmer suspects the mean mass of his turkeys has come down.

    (a) In a hypothesis test, what should be used as the test statistic?

    (b) Write down suitable null and alternative hypotheses.

    (c) State the type of test being carried out.

11.3 Summary

You have learnt the important language of statistical hypothesis testing.

You have learnt how to form two hypotheses: a **null hypothesis**, which represents no change, and an **alternative hypothesis**, describing a possible change in some **parameter**.

When performing a hypothesis test we form two hypotheses:

- A null hypothesis, denoted H_0, which we assume to be correct unless we find evidence to reject it;

- An alternative hypothesis, denoted H_1, which we will accept if we reject H_0.

Evidence is gained by taking a **sample**. The data from this sample is then summarised as a statistic called a **test statistic**. From this test statistic we can make an inference about the population.

The significance level of the test is the probability of incorrectly rejecting the null hypothesis.

The null hypothesis is never proven or disproved. Instead, we decide whether there is sufficient evidence to reject it, and therefore adopt the alternative hypothesis instead.

Chapter 12
Hypothesis Testing for the Mean in a Normal Distribution

12.1 Introduction

In natural processes, a random variable often follows a normal distribution; for example the masses of tree frogs in the Amazon, or the lengths of the worms in a back garden.

This chapter introduces hypothesis testing for such normally distributed random variables. It allows us to determine whether a change in the mean is likely to have been the result of some change in circumstances.

Key words and notation

In addition to the keywords covered in chapter 11, you will be introduced to these terms:

- **Population standard deviation**, σ: As with the population mean, this is a theoretical value, a parameter.
- **Population variance**, σ^2: The square of the population standard deviation.
- **Test statistic**, z_{test}: In the type of test covered in this chapter, the notation z_{test} is used for the test statistic.

In addition to the above, you will encounter the following notation:

- $\hat{\sigma}$ and $\hat{\sigma}^2$: Estimates of the population standard deviation and the population variance. (The ^ symbol is used to denote an estimate.)

Before you start

You should:

- Know how to find probabilities for a normally distributed random variable.
- Understand the terms: **null hypothesis**, **alternative hypothesis**, **test statistic**.
- Have gained an understanding of the process involved in testing hypotheses.
- Understand when to use a one-tailed or two-tailed test.

Exercise 12A (Revision)

1. Given: $X \sim N(35, 7^2)$ find $P(X < 30)$

2. Explain briefly the role of the sample and the test statistic in a hypothesis test.

3. To test whether the mean size of tree frogs in the Amazon has changed because of climate change, would a one-tailed test or a two-tailed test be required? Explain your answer briefly.

What you will learn

In this chapter you will learn how to:

- Conduct a statistical hypothesis test for the mean of a normal distribution with a known, given or assumed variance and interpret the results in context.

In the real world...

In 2012, scientists at the Large Hadron Collider (LHC), deep under the French and Swiss Alps, cleared up a 40-year-old mystery. They discovered the subatomic particle known as the Higgs Boson, sometimes known as the 'God Particle', which gives everything in the universe its mass.

To conclude that a new particle really has been found, particle physicists require that two independent particle detectors each conclude that there is less than a one-in-a-million chance that their observations are due to just background random events.

This is equivalent to saying that the observed number of events is more than five standard deviations (five sigma) different from that expected if there was no new particle.

On 4 July 2012 the physicists working on two independent experiments at the LHC announced they had independently made the same discovery. Both experiments independently reached the 5 sigma level, implying that the probability of this happening by chance alone was less than one in three million. The Higgs Boson had finally been found.

12.2 Sample Mean

Real-life data is subject to natural variation. This is something that needs to be taken into consideration when carrying out a hypothesis test. The measures of the degree of variation are the standard deviation and variance.

Consider taking a sample of size n from a population and calculating the mean, $\bar{X}$. If we repeat the process, a different mean may be found. Therefore, this **sample mean** can be considered a random variable.

Imagine repeating this process, always taking samples of size n, so that we obtain many instances of this random variable $\bar{X}$. An important result is: if the distribution of the population is normal with a mean μ and a variance σ^2, then the distribution of the sample mean will also be normal with mean μ and variance $\dfrac{\sigma^2}{n}$.

> **Note:** In other words:
>
> if $X \sim N(\mu, \sigma^2)$
>
> then $\bar{X} \sim N\left(\mu, \dfrac{\sigma^2}{n}\right)$
>
> where n is the size of the samples.

This makes intuitive sense:

For the mean: If we repeatedly take samples, we might expect the sample mean to have the same mean as the underlying population.

For the variance: For larger samples, we should expect less variation in the sample mean. Hence the variance is lower.

Worked Example

1. There are 10 000 people in a football stadium. The distribution of their ages closely approximates a normal distribution with a mean of 40 and a standard deviation of 5.
   (a) Write down the distribution of the population.
   (b) Consider repeatedly taking samples of the people in the stadium, with a sample size of:
   (i) 100   (ii) 10
   What is the distribution of the sample mean in each case?

$\mu = 40$ and $\sigma = 5$

(a) For the distribution of the population,
$$X \sim N(40, 5^2)$$
$$\text{or } X \sim N(40, 25)$$

(b) (i) For the distribution of the sample mean with a sample size of 100:
$$\bar{X} \sim N\left(40, \frac{5^2}{100}\right)$$
$$\bar{X} \sim N(40, 0.25)$$

(ii) For the distribution of the sample mean with a sample size of 10:
$$\bar{X} \sim N\left(40, \frac{5^2}{10}\right)$$
$$\bar{X} \sim N(40, 2.5)$$

The sample mean with the higher sample size shows much less variability.

Exercise 12B

1. (a) Explain what is meant by the sample mean.
   (b) Is the sample mean always the same as the population mean?

2. A population is distributed normally with a mean of 400 and a standard deviation of 20.
   (a) State the distribution the sample mean follows, given a sample size of 10.
   (b) How does the distribution of the sample mean change if the sample size increases?

3. There are 500 guests in a hotel. Their ages approximately follow a normal distribution with a mean of 35 and a variance of 42.25.
   (a) Write down the distribution of the ages, X.
   (b) Every morning a random sample of 10 guests is chosen and these guests receive a free breakfast. What is the distribution of the sample mean?
   (c) What modelling assumptions have been made in answering part (b)?

4. There are 5000 fish in an aquarium. The distribution of their masses closely approximates a normal distribution with a mean of 550 grams and a standard deviation of 45 grams.
   (a) Write down the distribution of the masses, M.

Exercise 12B...

(b) Consider repeatedly taking samples of the fish in the aquarium, with a sample size of:
  (i) 10
  (ii) 100
  What is the distribution of the sample mean in each case?

12.3 Statistical Hypothesis Testing For The Mean Of A Normal Distribution

In the previous section you learnt that the sample mean approximately follows a normal distribution, with the same mean as the population and a standard deviation of $\frac{\sigma}{\sqrt{n}}$. That is: $\bar{X} \sim N\left(\mu, \frac{\sigma^2}{n}\right)$

In this section you will learn how to carry out a hypothesis test on the mean of a normally distributed random variable. The standard deviation σ will be known, given or assumed.

This involves consideration of the quantity $\frac{\bar{x} - \mu}{\sqrt{\sigma^2/n}}$

which can be thought of as a standardised sample mean. Given certain criteria are met (outlined below), this follows $N(0, 1^2)$.

As such, the standard normal distribution tables can be used to determine whether the sample mean is taking extreme values.

In this way, we can take a sample to test whether the mean of a population is significantly different from some given or supposed value.

This type of test is sometimes referred to as a z-test, since we are calculating, as our test statistic, a value of z from the standard normal distribution.

We can use a z-test when:

- We can make an assumption that the data in the underlying population are normally distributed; and
- The sample size is large enough ($n > 30$).

Note: There is also a requirement that the standard deviation σ is known. At the time of publication, the CCEA specification for A2 Applied Mathematics states that the standard deviation will be known, given or implied.

You will also learn how to interpret the results of z-tests in the context of the question.

Population standard deviation and variance

Either the population standard deviation σ or the population variance σ^2 should be available from the information you are given.

You may see the notation $\hat{\sigma}^2$. The ^ symbol denotes an estimate, so $\hat{\sigma}^2$ denotes an estimate of the population variance.

The test statistic z_{test}

For a normally distributed random variable, the test statistic is z_{test}.

It is calculated as follows:

$$z_{test} = \frac{\bar{x} - \mu}{\sqrt{\sigma^2/n}}$$

where:

- $\bar{x}$ is the sample mean
- μ is the population mean, assuming H_0
- σ^2 is the population variance
- n is the sample size.

The higher the value of $|z_{test}|$, the more unlikely it is that this event could happen under H_0.

You may see equivalent variations of this formula,

for example $z_{test} = \frac{\bar{x} - \mu}{\sqrt{\hat{\sigma}^2/n}}$ and $z_{test} = \frac{\bar{x} - \mu}{\sigma/\sqrt{n}}$

The second one of these formulae involves σ, the standard deviation. Look at the wording of a question carefully to make sure you are using standard deviation σ or variance σ^2 as appropriate.

....

Worked Example

2. The mass of a chocolate bar is advertised as 200 g. Its standard deviation is known to be 8 g. Gavin thinks that the chocolate bars are smaller than they used to be. He samples 150 bars to test his suspicion. He finds the mean mass of the bars he samples to be 195 g.
  (a) Assuming the mass of the chocolate bars is normally distributed, find the value of the test statistic z_{test}
  (b) Gavin's girlfriend asks him if it was necessary to sample 150 bars of chocolate. How can he justify this mathematically?

(a) $z_{test} = \frac{\bar{x} - \mu}{\sqrt{\sigma^2/n}}$

$z_{test} = \frac{195 - 200}{\sqrt{8^2/150}} = -7.65$

(b) The larger the sample, the more reliable the results are likely to be.

The critical z value z_{crit}

The test statistic z_{test} is compared with a critical z value z_{crit} to determine whether the null hypothesis is rejected.

The table below shows the values of z_{crit}, given to 3 decimal places, for the most common significance levels used.

Note that a 95% **confidence level** is equivalent to a 5% significance level.

You can also find these values from your calculator and from the normal probability tables.

Significance level α	Confidence level	z_{crit} for a one-tailed test (negative for left-sided; positive for right-sided)	z_{crit} for a two-tailed test
1% or 0.01	99% or 0.99	±2.326	±2.576
2% or 0.02	98% or 0.98	±2.054	±2.326
2.5% or 0.025	97.5% or 0.975	±1.960	±2.241
5% or 0.05	95% or 0.95	±1.645	±1.960
10% or 0.1	90% or 0.9	±1.282	±1.645

The comparison

The larger the value of $|z_{test}|$, the less likely this scenario could occur under the null hypothesis.

So, the null hypothesis is rejected if $|z_{test}| > |z_{crit}|$.

The process

The entire process can be summarised as follows:

1. Define the null and alternative hypotheses.
2. Define the population distribution.
3. Define the distribution of the sample means.
4. Calculate the value of z_{test}.
5. Either
   - calculate the p value associated with this value of z_{test}; or
   - find the critical z value z_{crit} from tables or calculator.
6. Compare the value of z_{test} with z_{crit}, or the p value with the significance level.
7. Accept or reject the null hypothesis with a concluding statement.

> **Note:** You may see step 3 being omitted from a given solution to a problem. In an exam setting, you can be awarded full marks without stating the sampling distribution. However, formally, this step is important, since the following step involves finding a test z value, which is based on the sampling distribution.

Worked Example

3. The label on a particular box of cereal states that it contains 800 grams. The variance of the masses is known to be 17 231 grams^2. Sue suspects the label is wrong and decides to investigate. She records the mass of the contents of fifty boxes and calculates the sample mean as 788.8 grams.
   (a) Carry out a suitable hypothesis test at a 5% significance level to see whether Sue's suspicion is correct.
   (b) How could the test be improved?

(a) The sample mean and variance are 788.8 and 17 231 respectively.

Form the hypotheses:
$H_0: \mu = 800$
$H_1: \mu \neq 800$

> **Note:** Remember to use **equality** in the null hypothesis and inequality (either >, < or ≠) in the alternative hypothesis. Also remember to use μ and not $\bar{x}$, since we are carrying out a test on the population mean.

This is a two-tailed test. Sue suspects the label is wrong, which means she suspects the given mass could be too high or too low.

Assuming H_0, then for the population:
$X \sim N(800, 17231)$

and for the sampling distribution:
$$\bar{X} \sim N\left(800, \frac{17231}{50}\right)$$

Calculate z_{test}:
$$z_{test} = \frac{\bar{x} - \mu}{\sqrt{\sigma^2/n}} = \frac{788.8 - 800}{\sqrt{17231/50}} = -0.603 \,(3 \text{ s.f.})$$

Find z_{crit} at the 5% level for a two-tailed test. The table above gives:
$z_{crit} = \pm 1.96$

The diagram of the standard normal distribution below shows that the z value of −0.603 does not lie within either of the critical regions.

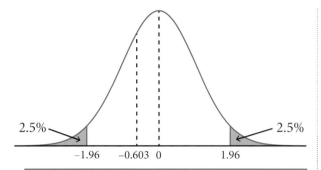

2.5%   2.5%

−1.96   −0.603 0   1.96

Note: You do not have to draw a diagram, but it may help you to visualise the problem.

Compare $|z_{test}|$ and $|z_{crit}|$.

Reject H_0 if $|z_{test}| > |z_{crit}|$:

$|z_{test}| = 0.603$ and $|z_{crit}| = 1.96$

Since $|z_{test}| < |z_{crit}|$ we do not reject H_0.

Give a conclusion in the context of the question:

There is insufficient evidence at the 5% significance level to suggest that the mean is not 800 g; therefore, we do not reject the null hypothesis.

Sue cannot conclude that the mean mass of the boxes of cereal is not 800 g.

(b) The test could be improved by using a larger sample.

Note: The bigger the sample, the more reliable the test is likely to be.

When performing a hypothesis test using a sample from a population that is distributed normally, be careful to use the correct significance level and the correct test: one-tailed or two-tailed.

Worked Examples

4. The vice-chancellor of a university claims that the mean salary of the university's graduates one year after graduation is £500 per week with a standard deviation of £75.

   A second-year student doubts the vice-chancellor's claim about the mean salary. He does a survey of 25 people who graduated one year ago, asking them their weekly salary. He finds the sample mean to be £460.

   Test at a 95% confidence level whether this is consistent with the vice chancellor's claim.

   We assume that the distribution of the salaries is normal:

$X \sim N(500, 75^2)$

Form the null and alternative hypotheses:

    $H_0: \mu = 500$

    $H_1: \mu \neq 500$

Note that a 95% confidence level is equivalent to a 5% significance level. From the table of critical z values above, we will reject H_0 if $|z_{test}| > 1.96$.

    $\mu = 500;\ \sigma = 75$

Since $X \sim N(500, 75^2)$:

$$\bar{X} \sim N\left(500, \frac{75^2}{25}\right)$$

$$\bar{X} \sim N(500, 15^2)$$

$$z_{test} = \frac{\bar{x} - \mu}{\sigma/\sqrt{n}} = \frac{460 - 500}{75/\sqrt{25}} = -2.67 \text{ (3 s.f.)}$$

For a two-tailed test with a 5% level of significance, $z_{crit} = \pm 1.96$.

    $|z_{test}| > |z_{crit}|$

This is also demonstrated by the following diagram, which shows that the z value of −2.67 lies inside the critical region.

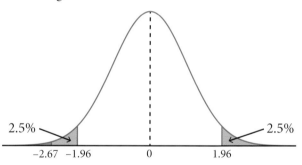

2.5%   2.5%

−2.67 −1.96   0   1.96

Therefore, the student rejects H_0 and adopts H_1.

There is evidence to suggest with 95% confidence that the vice chancellor's claim is incorrect.

Note: The student is checking whether the claim (that the mean is £500) is correct, i.e. checking whether it may be significantly higher or significantly lower than the £500 figure. Therefore, this is a two-tailed test.

If the wording in the question had been something like this:

A second-year student suspects the vice chancellor is exaggerating the average salary and suggests the true mean is lower,

then the student would need to conduct a left-sided one-tailed test, with the alternative hypothesis $H_1: \mu < 500$.

5. A car company is looking for fuel additives that might increase mileage. Without additives, their cars are known to average 25 mpg (miles per gallon) with a standard deviation of 2.4 mpg on a road trip from Belfast to Cork.

The company now asks whether a new additive increases the average mileage. In a study, thirty cars are sent on a road trip from Belfast to Cork. The 30 cars averaged $\bar{x} = 25.5$ mpg with the additive.

Can we conclude from this result that the additive is effective?

We are asked whether the new additive increases the mean miles per gallon. The current mean $\mu = 25$. We assume that the distribution of the car mileages is normal:

$$X \sim N(25, 2.4^2)$$

The null hypothesis is that nothing changes:

$$H_0: \mu = 25$$

The alternative hypothesis is that the mean increases:

$$H_1: \mu > 25$$

Now we need to calculate the test statistic. We start with the assumption the normal distribution still applies. Under the null hypothesis there is no change in μ.

The standard deviation $\sigma = 2.4$ mpg is known. We find z_{test} as follows:

$$z_{test} = \frac{\bar{x} - \mu}{\sigma/\sqrt{n}}$$

$$= \frac{25.5 - 25}{2.4/\sqrt{30}}$$

$$= 1.141$$

We are using a 5% significance level and a (right-sided) one-tailed test. From the tables we obtain $z_{crit} = 1.645$.

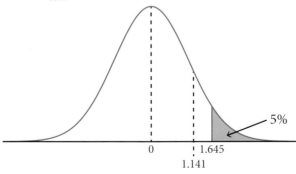

Since $|z_{test}| < |z_{crit}|$, as shown in the diagram, the test statistic is not in the critical region, and we cannot reject H_0.

The observed sample mean $\bar{x} = 25.5$ is consistent with the hypothesis $H_0: \mu = 25$ at a 5% significance level.

The following example demonstrates the use of the p value method.

We calculate z_{test} as in the previous examples, then find the probability of this occurring under the null hypothesis. For a one-tailed test, the probability is then compared with the significance level α, or with $\frac{\alpha}{2}$ for a two-tailed test.

Worked Example

6. The company that makes Crinkly Chips states that the masses of the bags of chips are normally distributed with a mean of 500 grams and a standard deviation of 10 grams.

Sinéad buys these chips regularly and starts to suspect that the mean mass of the bags is less than 500 grams. She decides to test her hypothesis using a significance level of 10%. She takes a random sample of 35 bags of chips and calculates the mean of the sample to be 496 grams.

Complete the hypothesis test.

$$H_0: \mu = 500; \ H_1: \mu < 500$$

This is a one-tailed test.

Mass is distributed normally with mean $\mu = 500$, $\sigma = 10$:

$$M \sim N(500, 10^2)$$

The sample mean follows the distribution:

$$\bar{M} \sim N\left(500, \frac{10^2}{35}\right)$$

$$z_{test} = \frac{496 - 500}{\sqrt{10^2/35}} = -2.366$$

Calculate the p value from the normal distribution tables or on the calculator:

$$p = P(\bar{M} < 496)$$

$$= P(Z < -2.366)$$

$$= 0.00899$$

The lower the probability, the less likely it is that a sample mean of 496 grams could have occurred under H_0.

The significance level α is 10% or 0.1.

Since $0.00899 < 0.1$ there is sufficient evidence at the 10% significance level to reject the null hypothesis and accept the alternative hypothesis.

There is sufficient evidence at the 10% significance level to suggest that the mean mass of the bags of chips is less than 500 g.

....

The next example demonstrates the p value method with a two-tailed test. In this situation the probability is compared with half of the significance level, to account for the possibility of being in either tail of the distribution.

....

Worked Example

7. A scientist reads in a physics journal that the mass of a recently-discovered particle is 900 MeV.

   She questions the methodology described in the article and therefore doubts the mass given. She decides to carry out an experiment and uses a hypothesis test with a 95% confidence level.

   After 20 runs of her experiment, she obtains a mean value of 908 MeV with a standard deviation of 12 MeV.
   (a) Form two hypotheses H_0 and H_1 for the test.
   (b) State whether the test is one-tailed or two-tailed.
   (c) Find the probability that the scientist's mean measurement of 908 MeV could occur under the null hypothesis.
   (d) Make a conclusion.
   (e) State one assumption the scientist has used in her hypothesis test.

   (a) $H_0: \mu = 900$; $H_1: \mu \neq 900$

   (b) This is a two-tailed test since the scientist doubts the given mean value, meaning it could be lower or higher.

   (c) Mass is distributed normally with mean $\mu = 900$, $\sigma = 12$:
   $$M \sim N(900, 12^2)$$

   The sample mean follows the distribution:
   $$\bar{M} \sim N\left(900, \frac{12^2}{20}\right)$$

   $$z_{test} = \frac{908 - 900}{\sqrt{12^2/20}} = 2.981$$

   $$p = P(\bar{M} > 908) = P(Z > 2.981) = 0.00144$$

   (d) Since this is a two-tailed test, compare p with $\frac{\alpha}{2}$.

The confidence level is 95%, which means that
$$\alpha = 0.05 \text{ and } \frac{\alpha}{2} = 0.025$$

$$0.00144 < 0.025$$

$p < \frac{\alpha}{2}$, which means it was unlikely this sample mean would be obtained under H_0. The scientist rejects the null hypothesis and accepts the alternative hypothesis. She concludes that there is evidence to suggest at the 5% significance level that the given mean of 900 MeV is wrong.

(e) The scientist is assuming a population standard deviation of 12 MeV, which is the sample standard deviation from her own experiment.

....

You have seen the two methods of comparison, which can be described as the z value method and the p value method.

In the following exercise, and in examination questions, you may use either method, unless the question specifically asks for one method.

Exercise 12C

1. In each part, a random sample of size n is taken from a population that follows a normal distribution with a mean μ and variance σ^2. Carry out a hypothesis test at the stated levels of significance.
   (a) $H_0: \mu = 20$; $H_1: \mu \neq 20$
   $n = 20$, $\bar{x} = 20.3$, $\sigma = 1.4$, at the 5% level
   (b) $H_0: \mu = 150$; $H_1: \mu < 150$
   $n = 35$, $\bar{x} = 140.3$, $\sigma = 10.4$ at the 10% level
   (c) $H_0: \mu = 8$; $H_1: \mu > 8$
   $n = 40$, $\bar{x} = 9.3$, $\sigma = 1.5$ at the 2.5% level
   (d) $H_0: \mu = 1000$; $H_1: \mu \neq 1000$
   $n = 60$, $\bar{x} = 1015$, $\sigma = 91$ at the 2% level

2. The national mean score in a general knowledge quiz is 14.7 with a standard deviation of 2.2. At Waterloo Road School, 50 students were selected at random. Thee mean of their scores was found to be 15.1.
   (a) Carry out a hypothesis test to determine, with 90% confidence, whether this school has a mean score that is higher than the national mean.
   (b) State one modelling assumption that has been made in answering part (a).

Exercise 12C...

3. A ferry with 1200 people on board capsizes and all the passengers end up in the sea. They are all rescued, but all spend different amounts of time in the water. The amount of time spent in the water is normally distributed.

   The captain of the ferry is later taken to court. A lawyer representing the passengers claims that the passengers spent a mean of 24 minutes in the water, with a standard deviation of 2 minutes.

   The captain disputes this claim and suggests the mean time was lower. He traces 50 of the passengers and asks them how long they were in the water. Amongst this group, the mean time was 20 minutes.

   (a) Write down suitable null and alternative hypotheses.
   (b) Is this a one-tailed or a two-tailed test? If one-tailed, is it left-sided or right-sided?
   (c) Find the value of the test statistic z_{test}.
   (d) State the critical value of z given a 5% significance level.
   (e) At a 5% significance level, can the court be confident that the mean time in the water was lower than 24 minutes?

4. Daily maximum temperatures are recorded in several locations across Northern Ireland. For the 30 years from 1961 to 1990, the summer maximum temperature, averaged over these locations, was 21.2°C, with a standard deviation of 5°C. In July 2021, the daily maximum temperature reached 31.3°C in Castlederg.

   Meteorologists wish to study whether temperatures have increased significantly. A researcher took a sample of 100 daily maximum temperatures from the summer months, randomly selected by location and date. The mean temperature in the sample is 22.1°C.

   (a) Write down suitable null and alternative hypotheses.
   (b) Is this a one-tailed or a two-tailed test?
   (c) Find the value of the test statistic z_{test}.
   (d) State the critical value of z given a 5% significance level.
   (e) At a 5% significance level, can the meteorologists claim they have found evidence that temperatures have increased?

Exercise 12C...

   (f) Do you think it is possible for the meteorologists to say they have found evidence of climate change? Justify your answer.

5. Patricia works in a lab testing water samples for bacteria. Over a long period of time, the mean concentration of bacteria in the River Lagan has been 950 per ml, with a standard deviation of 100 per ml.

   There was a storm at the weekend, and much water drained into the river. This week 50 water samples are taken along the river. The mean concentration of bacteria in these samples is 1000 per ml.

   Patricia suspects that the weekend storm has affected the concentration of bacteria in the water. She carries out a two-tailed hypothesis test at the 5% significance level, using the following hypotheses:

   $H_0: \mu = 950$ $\qquad$ $H_1: \mu \neq 950$

   Conduct the hypothesis test and state Patricia's conclusions.

6. In a computer game, a player's score depends heavily upon their reaction time. Scores are known to be normally distributed with a mean of 200 and a standard deviation of 30.

   Nuala wishes to test her theory that drinking coffee before playing the game affects your score. She selects a random sample of 120 adult volunteers and gives each a cup of coffee before they play the game. Nuala then calculates the mean score for this sample to be 205.

   (a) Nuala should carry out a hypothesis test. Explain briefly why this test should be two-tailed rather than one-tailed.
   (b) Test Nuala's theory using a 5% level of significance and make a conclusion.
   (c) What is the probability Nuala incorrectly rejects the null hypothesis?

7. In a fizzy drink factory, observations of the production line are taken to calculate the time taken to produce one pallet of drinks. The observations show that the time taken is normally distributed with a mean of 8 minutes and a variance 4.41 minutes².

Exercise 12C...

As a result of the introduction of a bonus scheme, the site supervisor claims that the workers are producing a pallet of drinks more quickly. It is found that, for a random sample of 170 inspections, the mean time taken is 7 minutes, 45 seconds. Test the supervisor's claim at the 5% significance level.

8. Over the last ten years, the times taken by the runners in a half-marathon have been normally distributed with a mean of 120 minutes and a standard deviation of 30 minutes. The organiser believes the mean time taken this year will be significantly different from the long-term mean. Following the race he selects the times of 50 runners at random. The mean time for this sample is 128 minutes.
   (a) Carry out a hypothesis test and show that the organiser's belief can be accepted at the 10% significance level.
   (b) Suggest one possible reason that the times this year differ from the long-term trend.

9. Peter suspects that the petrol pumps at his local petrol station are calibrated wrongly. He always buys 50 litres of fuel, but suspects he is actually getting less. Peter complains to the petrol station manager.

   The manager takes 40 samples of fuel, each of 50 litres according to the petrol pump display. He asks a laboratory to measure exactly how much fuel is in each sample. The lab reports that the average amount of fuel in the samples is 49.97 litres, and the sample standard deviation is 0.5 litres.
   (a) Assuming the samples follow a normal distribution, carry out a test at the 5% significance level to decide whether Peter's suspicions are correct.
   (b) What other assumptions have been made in answering part (a)?

10. Over a period of ten years, Chris has bought 700 items on an auction website. He determines the prices of these items closely follow a normal distribution, with a mean price of £9.60 and a standard deviation of £1.20. Chris believes that the prices of items on the site have increased over the last year. He samples, at

Exercise 12C...

random, 30 items he has bought during the last year. The mean price of these items is £10.20.
   (a) Test Chris's belief at a 99% level of confidence.
   (b) Comment on one possible way in which Chris's test may be flawed.

12.4 Summary

In this chapter you have learnt about the distribution of the sampling mean. Its mean μ is the same as that of the population. Its standard deviation is $\sigma/\sqrt{n}$, where σ is the population standard deviation and n is the sample size.

You have also learnt how to carry out a hypothesis test on a normally distributed variable.

In the CCEA specification for A2 Applied Mathematics (at the time of publication) either the standard deviation or the variance of the population will be known, given or implied.

Follow these steps:

1. State the distribution for the population, for example $X \sim N(20, 2^2)$.

2. State the sampling distribution, for example
$$\bar{X} \sim N\left(20, \frac{2^2}{30}\right).$$

3. Form the null and alternative hypotheses. The null hypothesis involves the population mean μ. It is a statement of what is currently true, for example $H_0: \mu = 20$

4. The alternative hypothesis is a statement describing the change we are testing for. For example $H_1: \mu > 20$

5. State whether the test is one-tailed or two-tailed. In the example above, the test is one-tailed, since we are only testing for an increase in the value of μ.

6. A sample is required and a test statistic z_{test} is calculated using the following formula:
$$z_{test} = \frac{\bar{x} - \mu}{\sqrt{\sigma^2/n}}$$

   where:
   • $\bar{x}$ is the sample mean,
   • μ is the population mean, assuming H_0,
   • σ^2 is the population variance, or an estimate of it,
   • n is the sample size.

7. Find the critical z value z_{crit} at the appropriate level of significance and for the appropriate type of test. This can be done using a table of values, on the calculator, or using the normal probability table on the formula sheet.

   Alternatively find the p value associated with your value of z_{test}.

8. Compare z_{test} with z_{crit}. If $|z_{\text{test}}| > |z_{\text{crit}}|$ then the outcome is unlikely under H_0, so H_0 is rejected and H_1 accepted. Otherwise H_0 cannot be rejected.

   Alternatively, compare the p value with the significance level α for a one-tailed test, or $\dfrac{\alpha}{2}$ for a two-tailed test.

   If $p < \alpha$ (for a one-tailed test) or $p < \dfrac{\alpha}{2}$ (for a two-tailed test), then reject H_0 and accept H_1. Otherwise do not reject H_0.

9. Give a conclusion. This should firstly state whether or not there is sufficient evidence to reject H_0 and accept H_1. Secondly, it should be phrased in the context of the question, for example 'There is insufficient/ sufficient evidence to conclude that the scores were higher.'

Chapter 13
Hypothesis Testing for The Proportion In A Binomial Distribution

13.1 Introduction

In chapter 11 you learnt the foundations of hypothesis testing, including the correct terminology and notation.

In chapter 12 you learnt how to carry out a hypothesis test to determine the significance of a mean value obtained from a sample in a normal distribution.

In this chapter you will apply hypothesis testing to discrete random variables that follow a binomial distribution. In this way you can determine whether the reported p value is accurate.

We often say we are conducting a hypothesis test for the **proportion** p, since p is the probability of success, but is also the proportion of successful outcomes.

Key words

- **Test statistic**, X: In the type of test covered in this chapter, the test statistic is a random variable X, which is the number of 'successes' in a binomial probability distribution.
- **Critical value**: In the type of test covered in this chapter, a critical value of X is used to determine whether a result is significant. It is the first value to fall inside the critical region.
- **Critical region**: Values of X for which H_0 is rejected.
- **Acceptance region**: Values of X for which H_0 is accepted.
- **p value**, p: The probability of the test statistic taking the observed value under the null hypothesis.
- **Actual significance level**: With a discrete distribution, such as the binomial distribution, the probability of the critical region is rarely equal to the significance level. The actual significance level is the combined probability of the critical region. It is usually slightly lower than the significance level.

Before you start

You should:

- Know when the binomial distribution can be used.
- Know how to calculate probabilities using the binomial distribution.
- Know how to calculate probabilities using the cumulative binomial distribution.
- Understand the notation and terminology used in hypothesis testing, including the terms null hypothesis, alternative hypothesis and test statistic.
- Have gained an understanding of the method involved in hypothesis testing.
- Understand when to use a one-tailed or two-tailed test.

Exercise 13A (Revision)

1. A production line makes chains. If a chain has one or more broken links, it is considered defective. The probability that a chain is defective is 0.08. A sample of 16 chains is taken. Find:
   (a) The probability that exactly two chains are defective.
   (b) The probability that no more than three chains are defective.

2. $X \sim B(10, 0.3)$. Find:
   (a) $P(X = 4)$
   (b) $P(X \geq 8)$
   (c) $P(2 \leq X \leq 5)$

3. A hypothesis test is carried out to determine whether the population of butterflies has changed during the last year. Should a one-tailed or a two-tailed test be carried out?

What you will learn

In this chapter you will learn:

- How to conduct statistical hypothesis tests for the proportion in the binomial distribution, including:
  - how to find critical regions and critical values;
  - how to conduct one-tailed and two-tailed tests; and
  - how to interpret the results of your hypothesis test in context.

In the real world...

In 1998, a construction company noticed a worrying trend in the type of accidents occurring on one of their major construction sites: accidents involving ladders.

All staff had been given the required training: the ladder should be at the right angle, so that it doesn't slip. It should be on firm ground. The worker should not climb too high and should not over-reach to one side of the ladder. He or she should also wear the correct protective clothing, in case an accident does occur.

But the type of accident being reported was not what the company management expected. As a worker descended and reached the second rung of the ladder, they often thought they had reached the bottom rung. They were then stepping out as if stepping onto the ground, leading to a trip or fall. Even a fall from such a low height can result in a back injury or a broken ankle.

The construction company introduced a new ladder that had many new safety features, including a textured bottom rung. Even through the heavy boots required on the site it was possible to detect a change in the texture as a worker reached the bottom rung. In this way, the ladder's manufacturer claimed, a worker is less likely to have this type of accident.

To test whether the introduction of this new ladder made a real difference to the number of accidents, a statistical hypothesis test was carried out, comparing the number of accidents during the year before the change and the number in the year after the change. In this particular test, the data showed a smaller number of accidents after the change, but the results were not significant at the required confidence level, so the company was not able to say with confidence that the introduction of the new ladder was making a significant change.

13.2 Critical Regions And Critical Values

The **critical region** is the set of possible values of the test statistic such that the combined probability of getting one of those values is **less than or equal to** the significance level.

The **critical value** is the first value to fall inside the critical region.

The **actual significance level** is the total probability of the observed value lying in the critical region.

....

Worked Example

1. A fair spinner has 5 sections: 2 of them yellow and 3 red.
   (a) Write down the probability p of getting a yellow when the spinner is spun.
   (b) Lynne adds a piece of sticky tape to one of the yellow sections. She thinks this makes it more likely the spinner will land on yellow. Write down a suitable null and alternative hypothesis Lynne can use to test this.

   Lynne spins the spinner 7 times and the number of yellows is counted.
   (c) Using a 5% significance level, find the **critical region** and the **acceptance region** for the hypothesis test.
   (d) Write down the **critical value**.
   (e) State the actual significance level of this test.

   _____

   (a) $p = \dfrac{2}{5} = 0.4$

   (b) $H_0: p = 0.4$; $H_1: p > 0.4$

   (c) Let X be the number of yellows out of the 7 outcomes in the sample. Then, assuming H_0 to be true, $X \sim B(7, 0.4)$.

   The following bar chart shows the probability distribution. You are not expected to draw a diagram like this, but it allows you to visualise the probabilities in this example.

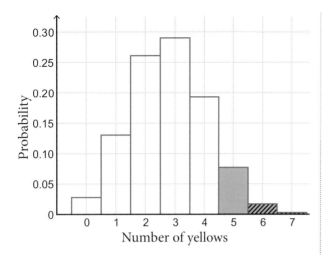

The height of each bar shows the probability of each corresponding X value. For example, the probability of getting three yellows is roughly 0.29. The blue shaded region shows the probability $P(X \geq 5)$. The area shown with hatching is $P(X \geq 6)$.

Using the binomial cumulative distribution table in the formula booklet, or using your calculator:
$$P(X \geq 5) = 1 - P(X \leq 4)$$
$$= 1 - 0.9037$$
$$= 0.0963$$

$$P(X \geq 6) = 1 - P(X \leq 5)$$
$$= 1 - 0.9812$$
$$= 0.0188$$

> **Note:** See section 13.4 for an example demonstrating how to find cumulative probabilities using the table. See section 13.5 for an example demonstrating how to do equivalent calculations on the calculator.

$P(X \geq 5) > 0.05$, whereas $P(X \geq 6) < 0.05$.

We look for the region that has a probability less than or equal to the significance level of 5%. Therefore, the critical region is $X = 6, 7$. The acceptance region is $0 \leq X \leq 5$.

(d) The critical value is the first value inside the critical region, so in this case it is 6.

(e) The actual significance level is the combined probability that, under H_0, X lies in the critical region. It is 0.0188 or 1.88%.

Read the question carefully. The following example states 'The probability in each tail should be as close as possible to 2.5%.' This changes the methodology slightly when finding the critical region.

....

Worked Example

2. An event has possible outcomes of success and failure. A long period of observations has shown that the probability of success $p = 0.2$. Some change occurs and a test is required for the following hypotheses:
   $H_0: p = 0.2$; $H_1: p > 0.2$
   A sample of 30 observations is taken and the proportion of successes is calculated.
   (a) Using a 5% significance level, find the critical region for this test. The probability in each tail should be as close as possible to 2.5%.
   (b) State the actual significance level of this test.

   ──────────────────────────────

   (a) Let X be the number of successes out of the 30 in the sample. Then, assuming H_0 to be true, $X \sim B(30, 0.2)$.

   For the left-hand tail (using the binomial cumulative distribution table):
   $$P(X \leq 1) = 0.0105$$
   $$P(X \leq 2) = 0.0442$$

   The lower critical region is the values 0 and 1, since 0.0105 is the closest value to 2.5%.

   For the right-hand tail:
   $$P(X \geq 11) = 1 - P(X \leq 10)$$
   $$= 1 - 0.9744$$
   $$= 0.0256$$

   $$P(X \geq 12) = 1 - P(X \leq 11)$$
   $$= 1 - 0.9905$$
   $$= 0.0095$$

   $P(X \geq 11)$ is slightly greater than 2.5% and $P(X \geq 12)$ is less than 2.5%.

   Ordinarily we would conclude that the upper critical region is the range of values from 12 to 30. But since the wording in the question requires us to use a probability in each tail as close as possible to 2.5%, we choose the values from 11 to 30.

   (b) The actual significance level is the combined probability that X lies in the critical region, i.e.
   $0.0105 + 0.0256 = 0.0361$

....

Exercise 13B

1. For each of the pairs of hypotheses below, state whether they describe a one-tailed or two-tailed test. If one-tailed, state whether the test is left-sided or right-sided.
   (a) $H_0: p = 0.4$; $H_1: p > 0.4$
   (b) $H_0: p = 0.1$; $H_1: p < 0.1$
   (c) $H_0: p = 0.75$; $H_1: p \neq 0.75$

2. Explain what you understand by the following terms:
   (a) critical region,
   (b) acceptance region,
   (c) critical value.

3. Debi carries out a hypothesis test. She rejects the null hypothesis and calculates the actual significance level of the test to be 0.0787. What is the probability she incorrectly rejects the null hypothesis?

4. An archery target has 10 rings. The centre two rings are gold in colour. Over a long period of time, Sam has worked out that, when she fires an arrow, the probability of her arrow landing in the gold zone is $p = 0.15$. Sam thinks her archery skills have improved.
   (a) Write down a suitable null and alternative hypothesis she can use to test this. State also whether this test is one-tailed or two-tailed.
   (b) Sam fires 20 arrows and the number that land in the gold zone is counted. Using a 5% significance level, find the critical region for the hypothesis test.
   (c) Write down the critical value.
   (d) State the actual significance level of Sam's test.

5. Every year Seamus sits a summer multiple choice maths exam. For each question there are four options. Last year, he guessed every answer.
   (a) For each question, what was the probability Seamus got it right?
   (b) Seamus tells his teacher his maths is now improving. Write down a null and alternative hypothesis the teacher can use to test this.
   (c) In this year's test there are 40 questions. Using a 5% significance level, find the critical region. The probability in the tail

Exercise 13B...

of the distribution should be as close as possible to 5%.
   (d) State the actual significance level of this test.

6. Every year Curtis plants 36 pea plants. On average, 12 pea plants survive to be planted on. This year he decides to use a new fertiliser on the plants which he believes will improve the number that survive.
   (a) Describe the test statistic and state suitable null and alternative hypotheses.
   (b) Using a 10% level of significance, find the critical region to check his belief.
   (c) State the probability of incorrectly rejecting the null hypothesis using this critical region.

7. A spinner has five sides numbered 1, 2, 3, 4 and 5. Hardeep thinks it is biased towards giving a 5. He spins the spinner 20 times and counts the number of times X that he gets a 5. Hardeep decides to do a hypothesis test with a level of significance of 5%.
   (a) Describe the test statistic X.
   (b) Write down suitable null and alternative hypotheses.
   (c) What values of X would cause the null hypothesis to be rejected?

13.3 Hypothesis Testing

This section discusses hypothesis testing on random variables that follow a binomial distribution.

You will learn some new terminology, and the examples will demonstrate how to interpret the results of the test in context.

We collect evidence in the form of a sample and the test statistic X is the number of 'successes'.

There are two ways to perform a hypothesis test for the proportion of a binomially distributed random variable:

- Finding the critical region, as in the previous examples. We then decide whether the outcome we observe lies within this region.

- Finding the probability of something happening under the null hypothesis and comparing this probability with the significance level. Applying the test, we determine whether the observed X value is likely to occur under the null hypothesis. If the probability of this number of successes is low under the null hypothesis, the null hypothesis is rejected. Otherwise, we cannot reject it.

Exam questions will often require you to find the critical region.

At the outset of the test, a significance level should be stated, which is often 5%, as with the tests covered in the previous chapter. The significance level is denoted by the Greek letter α, so in this case $\alpha = 0.05$.

For a two-tailed test, the critical region consists of two parts, one at each end of the distribution, as demonstrated in the next example.

··

Worked Example

3. A nutritionist is conducting an experiment into the effects of diet on memory. She presents a volunteer with a tray of 10 objects and lets the volunteer look at them for one minute before taking the tray away. She then asks the volunteer to write down as many of the objects as possible.

   This experiment is repeated each day for a week, after which the nutritionist determines that the proportion p of objects the volunteer can remember is 0.4.

   During the next phase of the experiment, the nutritionist puts the volunteer on a diet including supplements. The experiment is repeated with a view to deciding whether this changes the number of objects the volunteer can remember.
   (a) Find the critical region for a two-tailed test using a 10% level of significance. State also the critical values in each tail.

   Following the change in diet, the volunteer recalls 6 out of 10 of the objects successfully.
   (b) What are the nutritionist's conclusions?
   (c) Find the **actual significance level** of this test.

   (a) This is a two-tailed test, since the question states the nutritionist is testing whether the supplements cause **changes** in the scores, not increases or decreases. Since the significance level is 10%, we take 5% in each tail.

   If X is the number of objects successfully recalled, then: $X \sim B(10, 0.4)$

   The null hypotheses assumes no change:
   $H_0: p = 0.4$
   $H_1: p \neq 0.4$

   Use the binomial cumulative distribution function in the formula booklet. For $n = 10$ and $p = 0.4$:

x	$P(X \leq x)$
0	0.0060
1	0.0464
2	0.1673
3	0.3823
4	0.6331
5	0.8338
6	0.9452
7	0.9877
8	0.9983
9	0.9999
10	1.0000

$x = 0$ and $x = 1$ are included in the critical region, since $P(X \leq 1) = 0.0464$, which is less than 0.05.

> **Note:** $P(X \leq 2) > 0.05$, so 2 is not included in the critical region.

$x = 8, x = 9$ and $x = 10$ are included in the critical region, since $P(X \geq 8) = 1 - P(X \leq 7) = 0.0123$ which is less than 0.05.

> **Note:** $P(X \geq 7) = 1 - P(X \leq 6) = 0.0548$ $0.0548 > 0.05$ so 7 is not included in the critical region.

Therefore, the critical region (highlighted in blue) is: $x = 0, 1, 8, 9, 10$

The critical values in the lower and upper tails are 1 and 8 respectively.

(b) The nutritionist does not have enough evidence to reject the null hypothesis, since 6 is not in the critical region.

(c) The actual significance level is the total probability of the outcomes in the critical region.

Actual significance $= P(X \leq 1) + P(X \geq 8)$
$= 0.0464 + 0.0123$
$= 0.0587$

Another way of describing this is the probability of incorrectly rejecting the null hypothesis.

··

In the next example, the probability method is used.

In this case, Susan throws a die several times and counts the number of sixes, which is 4. Susan determines there is only a small probability of this result occurring under the null hypothesis, so she rejects the null hypothesis.

··

Worked Example

4. Susan wants to decide whether a six-sided die is biased towards landing on a 6. She rolls the die 8 times and obtains 4 sixes.

   (a) Write down:

   (i) the test statistic;

   (ii) a suitable null hypothesis; and

   (iii) a suitable alternative hypothesis.

   (b) Assuming the die is not biased, what is the probability of getting 4 or more sixes from 8 rolls of the die?

   (c) If the significance level is 5%, do we count this result as significant? What are Susan's conclusions?

   (d) In what way could Susan improve her test?

(a) (i) The test statistic is X, the random variable representing the number of sixes thrown on a standard die.

   (ii) Let p be the probability of the die landing on a 6. The null hypothesis is:

   $H_0: p = \dfrac{1}{6}$

   (iii) The alternative hypothesis is:

   $H_1: p > \dfrac{1}{6}$

(b) If X is the random variable representing the number of sixes thrown on a standard die, then:

$$X \sim B\left(8, \dfrac{1}{6}\right)$$

We require $P(X \geq 4)$. It is not possible to find this from the cumulative binomial distribution tables, since $p = \dfrac{1}{6}$ is not one of the listed probabilities.

Method 1

Work out the probability for each relevant value of X:

$$P(X = 4) = {}^{8}C_4 \left(\frac{1}{6}\right)^4 \left(\frac{5}{6}\right)^4 = 0.02605$$

$$P(X = 5) = {}^{8}C_5 \left(\frac{1}{6}\right)^5 \left(\frac{5}{6}\right)^3 = 0.004168$$

$$P(X = 6) = {}^{8}C_6 \left(\frac{1}{6}\right)^6 \left(\frac{5}{6}\right)^2 = 0.000417$$

$$P(X = 7) = {}^{8}C_7 \left(\frac{1}{6}\right)^7 \left(\frac{5}{6}\right)^1 = 2.4 \times 10^{-5}$$

$$P(X = 8) = {}^{8}C_8 \left(\frac{1}{6}\right)^8 \left(\frac{5}{6}\right)^0 = 6.0 \times 10^{-7}$$

Note: Alternatively, these probabilities can be worked out using Binomial Probability Distribution mode on your calculator, where you can enter a list of x values (4 to 8 in this case) to find many probabilities at once. See section 13.5 for instructions on calculator methodologies.

$$P(X \geq 4) = 0.02605 + 0.004168 + 0.000417$$
$$+ 2.4 \times 10^{-5} + 6.0 \times 10^{-7}$$
$$= 0.03066 \text{ or } 3.07\% \text{ (3 s.f.)}$$

Method 2

$P(X \geq 4)$ can be found using the Binomial Cumulative Distribution mode on the calculator. See section 13.5 for instructions on calculator methodologies.

$$P(X \geq 4) = 1 - P(X \leq 3)$$

Using the Binomial Cumulative Distribution mode with $x = 3$, $N = 8$ and $p = \dfrac{1}{6}$, we find:

$$P(X \leq 3) = 0.96934 \ldots$$

$$P(X \geq 4) = 1 - 0.96934 \ldots$$
$$= 0.03066 \text{ or } 3.07\% \text{ (3 s.f.)}$$

If it were a standard die (hypothesis H_0), there would be only a 3.07% chance of getting 4 or more sixes.

(c) This is a significant result, since $3.07\% < 5\%$. Susan concludes that there is evidence to reject H_0. In context, there is evidence to suggest that the die is biased towards getting a 6.

(d) Susan could increase the number of times she rolls the die. A larger sample usually gives better results.

···

In the previous example, Susan concludes that the die is biased, and she rejects the null hypothesis. However, she could be wrong. If her die were unbiased, she could still get 4 or more sixes from 8 rolls. The probability of this happening is 3.07%.

There is always a chance of rejecting the null hypothesis incorrectly.

Note: In a discrete probability distribution, such as the binomial distribution, the probability of incorrectly rejecting the null hypothesis is the **actual significance level**.

If a question does not specifically ask for the critical regions, either of the two methods outlined above can be used.

Worked Example

5. A TV consumer programme claims that 20% of a supermarket's carrier bags fall apart when the load they carry is 5 kg or more. The supermarket claims that the true percentage is lower and carries out a hypothesis test on its own bags. A random sample of 40 of the supermarket bags finds that just one of them falls apart when carrying a load of 5 kg. The results are analysed using a hypothesis test with a 5% significance level.
   (a) Write down suitable null and alternative hypotheses.
   (b) Find the critical region.
   (c) Test at a 5% significance level whether the supermarket can claim, on this evidence, that less than 20% of the bags fall apart.

(a) Our hypotheses are:
$H_0: p = 0.2$ $H_1: p < 0.2$

(b) Let X be the random variable representing the number of carrier bags that fall apart under a load of 5 kg. Then under H_0, $X \sim B(40, 0.2)$.

Under H_0, $P(X \leq 3) = 0.0285$. In other words, there is a 2.85% chance that 3 or fewer bags fall apart under H_0.

Whereas $P(X \leq 4) = 0.0759$, so there is a 7.59% chance that 4 or fewer bags fall apart.

The critical region is 0, 1, 2 and 3, since this is the region whose cumulative probability is less than the significance level of 5%.

(c) Since $X = 1$, this value falls within the critical region. Therefore, there is sufficient evidence to reject H_0 and adopt H_1.

There is sufficient evidence that the proportion is less than 0.2, or in other words insufficient evidence that 20% of the carrier bags fall apart with loads of 5 kg.

....

In the following example, the test statistic is calculated and the probability of this value occurring is compared with the significance level of the test.

....

Worked Example

6. The standard drug for treatment of a disease has a $\frac{1}{5}$ chance of success.

A new drug is introduced. The manufacturer claims that the new drug has a higher success rate. Following treatment with the new drug, 40 patients are followed up and it is found that for 9 of them, the new drug was successful. Test, at a 5% significance level, whether the new drug has increased the proportion of patients being treated successfully.

The test statistic is the number of patients X treated successfully. This is a random variable with a binomial distribution:

$$X \sim B\left(40, \frac{1}{5}\right)$$

The null hypothesis is that the new drug makes no difference:

$$H_0: p = \frac{1}{5}$$

The alternative hypothesis is that the new drug increases the proportion of patients treated successfully:

$$H_1: p > \frac{1}{5}$$

We calculate the probability that $X \geq 9$ under the null hypothesis:

$$P(X \geq 9) = 1 - P(X \leq 8)$$

From the binomial cumulative distribution table, when $n = 40$ and $p = 0.2$:

$$P(X \leq 8) = 0.5931$$
$$P(X \geq 9) = 1 - 0.5931$$
$$= 0.4069$$
$$= 0.407 \text{ or } 40.7\% \text{ (3 s.f.)}$$

Under the null hypothesis (in other words, if the new drug makes no difference), there is a 40.7% chance of 9 or more patients out of 40 being successfully treated. This is quite high.

$$40.7\% > 5\%$$

At a 5% significance level, there is not enough evidence to reject H_0.

In context, there is not enough evidence to suggest that the new drug is better than the old one.

....

Exercise 13C

1. For each set of conditions below, carry out a hypothesis test to determine whether or not H_0 should be rejected. n is the number of trials; x is the number of successful outcomes; α is the significance level and p is the proportion of successful trials expected under the null hypothesis H_0.

Exercise 13C...

(a) $H_0: p = \frac{1}{3}$; $H_1: p < \frac{1}{3}$; $n = 10$; $x = 3$; $\alpha = 0.05$

(b) $H_0: p = 0.7$; $H_1: p > 0.7$; $n = 20$; $x = 15$; $\alpha = 0.05$

(c) $H_0: p = 0.25$; $H_1: p \neq 0.25$; $n = 30$; $x = 2$; $\alpha = 0.05$

2. A traffic survey in Belfast suggests that 35% of cars driving in the city do not meet required emissions standards. A city councillor suggests this is not the case. She conducts a random sample of 15 cars to test the hypothesis at the 10% level.
   (a) Should the councillor's test be one-tailed or two-tailed? Justify your answer.
   (b) Write down suitable null and alternative hypotheses.
   (c) Determine the critical region, i.e. the values of X for which the null hypothesis should be rejected.
   (d) Find the corresponding actual significance level.

3. The owner of a women's football team is disappointed with recent results. Since the beginning of the season the team has won only 7 games out of 35. The owner tells the manager to improve results, or she will be fired. The manager makes a change to the team's tactics. Over the next 10 games, the team wins 3 games. The owner conducts a hypothesis test at a 10% significance level to decide whether or not to fire the manager. What does the owner conclude? Show all your working.

4. Finn tosses a coin twenty times. It lands on heads six times. He would like to know whether the coin is biased towards tails.
   (a) Should Finn conduct a one-tailed or two-tailed hypothesis test?
   (b) Write down the null and alternative hypotheses.
   (c) Perform a hypothesis test at a 5% significance level to see if the coin is biased and write down Finn's conclusion.

5. After a particular operation it has been shown that 20% of patients across Northern Ireland will require follow-up treatment. St Joseph's Hospital takes a sample of size 20 from the patients on their records who have had this operation.

Exercise 13C...

(a) Find the critical region at the 5% level of significance, to test whether the proportion of St Joseph's patients requiring follow-up treatment differs from the Northern Ireland proportion. The probability in each tail should be as close as possible to 2.5%.

(b) State the actual significance level of the test.

(c) The hospital finds that 8 out of their 20 patients require follow up treatment. Comment on this finding in light of your critical region.

6. It has been established that only 30% of students on a particular university course reach the required level by the end of their first year. Ten students are chosen at random from those starting their first year.
   (a) Calculate the probability that at least one of the students has reached the required level by the end of the year.
   (b) The head of department introduces a new structure to the course, which he believes will improve the proportion of students reaching the required level at the end of the first year. He selects 10 students at random who are starting the course. At the end of the year 6 of them have reached the required level. Perform a suitable hypothesis test to determine, at the 5% level, whether the head of department's belief is true.
   (c) State how the hypothesis test could be improved.

7. In a bakery, various loaves of bread are on sale. Over a long period of time, the manager works out that, of those customers buying bread, one in ten buys the wholegrain loaf. This week, a sample of 50 customers who bought bread was taken. Eleven of them chose the wholegrain loaf. The manager thinks this is evidence the wholegrain loaf has become more popular this week. Test the manager's belief at a 1% significance level.

8. In a bag of Moonburst sweets there are five different flavours. Until now, there have been the same number of each flavour. Over the last month, Charlie thinks the green lime sweets have become more common in each bag. He takes a sample of 30 sweets to test his hypothesis and finds 8 lime sweets.

Exercise 13C...

(a) Write down the null and alternative hypotheses that Charlie uses.

(b) If X is the random variable 'the number of lime sweets in the sample', what are the critical values for X, assuming a significance level of 5%?

(c) Complete the test and make a conclusion.

(d) What is the actual significance level of this test?

9. A machine makes T-shirts and it is observed that one in 10 of the T-shirts have defects in the stitching. The production process is modified and a sample of 20 T-shirts is taken. One of the T-shirts has defects. Test, at the 10% level of significance, the hypothesis that the proportion of T-shirts with defects has changed because of the change in the production process. State your hypotheses clearly.

10. Kate's apple tree sometimes gives her bad apples, although mostly they are blemish-free and delicious. Last year Kate had a bumper crop of 160 apples, of which 56 were bad. This year Kate suspects the proportion of apples that are bad will be different. She takes an initial sample of 12 apples, half of which are bad. She conducts a hypothesis test, using a significance level of 10%.

(a) Does Kate carry out a one-tailed or two-tailed test?

(b) Write down Kate's null and alternative hypotheses.

(c) Find the critical region for the number of bad apples.

(d) Complete the test and make a conclusion.

(e) What is the actual significance level of this test?

(f) What could Kate do to improve her test?

11. Opinion polls in the United Kingdom regularly ask this question:

Do you think Britain should continue to have a monarchy in the future, or should it be replaced with an elected head of state?

There are three possible responses: (1) Support the monarchy, (2) Support for an elected head of state, (3) Don't know / don't mind. Over many years, support for the monarchy has been consistent, at around 62%.

Exercise 13C...

Freya believes that the death of Queen Elizabeth II in 2022 has changed attitudes. Shortly after the Queen's funeral, she conducts her own opinion poll of 40 people, selected at random, asking them the same question. She found that 28 people out of 40 responded 'Support the monarchy'.

Carry out a hypothesis test at a 5% level to determine whether Freya's belief is correct.

12. A YouTuber is desperate for fame. Over the last three years, however, surveys have revealed that only one in 100 people have heard of him.

Last month the YouTuber released a video that went viral. He is now sure that the proportion of people who have heard of him has increased.

He carries out a survey of 40 people. Two of them knew the YouTuber's name. Test his belief at the 5% level of significance, ensuring that the probability in the distribution's tail is as close as possible to the significance level.

13.4 How To Use The Binomial Cumulative Distribution Table

You can use the Binomial Cumulative Distribution Function table in your formula booklet.

Given that $X \sim B(n, p)$, the table gives $P(X \leq x)$ for a range of n values from 5 to 40, a range of p values between 0 and 0.5 and x values from 0 to n.

Worked Example

7. $X \sim B(6, 0.15)$
Find to 4 decimal places:
(a) $P(X \leq 3)$
(b) $P(X \geq 5)$

(a) The table shows that, for $n = 6, p = 0.15$ and $x = 3$:
$P(X \leq 3) = 0.9941$

$p =$	0.05	0.10	0.15	0.20
$n = 5, x = 0$	0.7738	0.5905	0.4437	0.3277
1	0.9774	0.9185	0.8352	0.7373
2	0.9988	0.9914	0.9734	0.9421
3	1.0000	0.9995	0.9978	0.9933
4	1.0000	1.0000	0.9999	0.9997
$n = 6, x = 0$	0.7351	0.5314	0.3771	0.2621
1	0.9672	0.8857	0.7765	0.6554
2	0.9978	0.9842	0.9527	0.9011
3	0.9999	0.9987	0.9941	0.9830
4	1.0000	0.9999	0.9996	0.9984
5	1.0000	1.0000	1.0000	0.9999

(b) Since the Binomial Distribution is a discrete distribution, only the values 0, 1, 2, 3, 4, 5 and 6 are possible. Therefore:

$$P(X \geq 5) = 1 - P(X \leq 4)$$
$$= 1 - 0.9996$$
$$= 0.0004$$

13.5 Calculator Methodologies

Some calculators have functions allowing calculation of the Binomial Probability Distribution $P(X = x)$ and the Binomial Cumulative Probability $P(X \leq x)$. The Binomial Cumulative Probability function is an alternative to using the table in the formula booklet.

The tips here are written for the Casio fx-991EX *Classwiz*, although other Casio models may have the same functionality and the instructions will be similar.

Binomial probability distribution $P(X = x)$
You can find probabilities relating to the binomial distribution on your calculator. To find $P(X = x)$ on your calculator, follow these steps:

1. Press **MENU**, then **7** for Distribution.
2. Press **4** for Binomial PD (Binomial Probability Distribution).
3. Press **1** for **list** or **2** for **variable**.
4. If using a list, enter all values of x for which you require a probability. Press = to enter each one, then = again to finish. Then enter values for N and P and press =.
   If using a single x value, enter values for x, N and P and press =.

Worked Example

8. Using the calculator, find the probability of 5 successes from 10 trials, with a probability of 0.2 of success in each trial.

We require $P(X = 5)$.

Enter Menu 7, Binomial PD, then 2 for variable.

Entering $x = 5$, $N = 10$ and $p = 0.2$ gives
$P = 0.0264 \ldots$

$P = 0.0264$ (3 s.f.)

Binomial cumulative distribution $P(X \leq x)$
You can find probabilities relating to the binomial cumulative distribution on your calculator. To find $P(X \leq x)$ on your calculator, follow these steps:

1. Press **MENU**, then **7** for Distribution.
2. Press the down arrow to get more options, then **1** for Binomial CD (Binomial Cumulative Distribution).
3. Press **1** for **list** or **2** for **variable**.
4. If using a list, enter all values of x for which you require a probability. Press = to enter each one, then = again to finish. Then enter values for N and P and press =.
   If using a single x value, enter values for x, N and P and press =.

Worked Example

9. Consider 10 trials that can result in success or failure. The probability of success in each trial is 0.3. Using the calculator, find the probability of:
   (a) no more than 5 successes from 10 trials,
   (b) no fewer than 8 successes from 10 trials.

(a) 'No more than 5' means 5 or fewer, so we require $P(X \leq 5)$.
   Enter Menu 7, Binomial CD, then 2 for variable.
   Entering $x = 5$, $N = 10$ and $p = 0.3$ gives:
   $$P = 0.95265 \ldots$$

   $P(X \leq 5) = 0.953$ (3 s.f.)

(b) 'No fewer than 8' means 8 or more, so we require $P(X \geq 8)$. This can be calculated as $1 - P(X \leq 7)$

   To find $P(X \leq 7)$: enter $x = 7$, $N = 10$ and $p = 0.3$. The calculator gives:
   P = 0.9984096 ...

   $P(X \leq 7) = 0.9984096 \ldots$
   $P(X \geq 8) = 1 - 0.9984096 \ldots = 0.00159$ (3 s.f.)

13.6 Summary

This chapter introduced the ways in which a hypothesis test can be carried out to test the proportion in a binomial distribution.

As with any hypothesis test, a sample is taken. The sample is then used to calculate the test statistic, which is the number of 'successes'.

Null and alternative hypotheses should be formulated, relating to the proportion p of successes.

There are two ways to carry out the test.

- Find the probability of this number of successes occurring under the null hypothesis. Then compare this probability with the significance level. A low probability of this occurrence leads to rejection of the null hypothesis.

- Find the critical region and decide whether the outcome observed lies within it. If so, the null hypothesis is rejected.

Exam questions will often require you to find the critical region.

Following completion of the test, a conclusion should be presented in the context of the question.

Chapter 14
Hypothesis Testing For Correlation

14.1 Introduction

In AS Applied Mathematics you learnt how to calculate the product-moment correlation coefficient r for **bivariate data**. You were introduced to the idea that an r value of more than 0.7 denotes a strong correlation between the variables.

In this chapter, you will learn a more sophisticated approach, involving hypothesis testing, to determine whether a correlation coefficient is significant.

Key words and notation

- **Bivariate data**: A dataset in which two variables are involved.
- **Correlation**: A link or dependence between two variables.
- **Test statistic**, r: In the type of test covered in this chapter, the test statistic is the product-moment correlation coefficient r.
- **Critical value**, r_{crit}: In the type of test covered in this chapter, r is compared with a critical r value to determine whether it is significant. The critical value is found in a table in the formula booklet.
- **Critical region**: Values of r for which H_0 is rejected.
- **Acceptance region**: Values of r for which H_0 is accepted.
- **p value**, p: The probability of finding the observed value or a more extreme value using the null hypothesis.
- **One-tailed test and two-tailed test**: A one-tailed test is used to determine whether a correlation exists in one direction (positive or negative). A two-tailed test is used to determine whether a correlation exists in either direction (either positive or negative).

Before you start

You should recall from AS Applied Mathematics:

- That the product-moment correlation coefficient is a measure of correlation between two variables.
- How to calculate and interpret the **product-moment correlation coefficient**.

You should also understand that:

- To perform a hypothesis test, you formulate two competing hypotheses: the null hypothesis and the alternative hypothesis.
- The null hypothesis is the default position that the effect you are looking for does not exist.
- The alternative hypothesis is that your prediction is correct.
- The goal of hypothesis testing is to collect evidence and reject the null hypothesis if it appears unlikely to be true. In other words, we reject the null hypothesis if there is some experimental support for the alternative hypothesis. Note that we have not **proved** the alternative hypothesis is true.

Exercise 14A (Revision)

1. Six competitors take part in show jumping trials. Each competitor is given a mark out of ten by two judges, Judge J and Judge K, as shown in the table.

Competitor	A	B	C	D	E	F
Judge J	8	6	8	7	4	9
Judge K	7	7	7	6	5	8

   (a) Find the product-moment correlation coefficient between the scores of the two judges.
   (b) Comment on the value obtained in part (a).

2. On Bernard's turkey farm it has been determined, over a long period of time, that the proportion p of turkeys reaching the required weight before Christmas is 0.75. This year Bernard thinks that proportion is significantly higher.
   (a) Write down two hypotheses Bernard can use to test his theory.

Exercise 14A...

(b) Should Bernard carry out a one-tailed or two-tailed test? Explain briefly why.

(c) Bernard decides to test at a significance level of 10%. His test shows that he should reject the null hypothesis. Write down the probability that he rejects the null hypothesis incorrectly.

(d) Having rejected the null hypothesis, Bernard releases a statement saying 'We have proved that more of our turkeys are larger this Christmas!' Comment on this statement.

What you will learn

In this chapter you will learn:

- How to interpret a given correlation coefficient using a given p-value or critical value. In other words, we are testing whether a correlation between two variables is significant.

In the real world...

In 1983, a TV documentary revealed a high number of childhood leukaemia cases between 1955 and 1983 in the village of Seascale, in north-west England, and it caused a public outcry. Seascale lies just a few miles from the Sellafield nuclear site. There were seven cases of leukaemia, where less than one would have been expected during this time.

Residents, politicians and the public were shocked and understandably worried. Unsurprisingly, some suggested that radioactive discharges from Sellafield may be to blame.

The documentary led to the setting up of COMARE, the Committee on Medical Aspects of Radiation in the Environment. It carried out investigations confirming that rates of both leukaemia and non-Hodgkin lymphoma were significantly higher than expected in Seascale.

The researchers found that the increased cancer rate wasn't mirrored in other nearby areas. They also found that the radiation dose from the site was dwarfed by the amount of radiation from natural sources such as radon gas and naturally occurring radioactivity in certain foods.

The experts also looked at the possibility that radiation the mother was exposed to before or during pregnancy was responsible for the cases and found no evidence that this was the case.

In 2016, COMARE's report was finally released. It confirmed that the cancer cluster had disappeared; and concluded that radiation wasn't to blame.

There was certainly correlation between the number of cases and their proximity to Sellafield. But the hypothesis that radiation from the site was responsible for the cancers was rejected.

Recent research points to two other factors: infection and population-mixing. While leukaemia itself is not infectious, infections may play a part in its development. Population-mixing occurs when relatively large numbers of people from urban areas move into rural communities, exposing local people to new infections. In the case of Seascale, an influx of workers moving there to work at the nuclear site may have played this role.

Similar effects had been observed during World War II, when soldiers and munitions workers were relocated in isolated communities. It also seems to have occurred around proposed nuclear sites where construction never actually took place.

While there is some evidence that infection and population-mixing may have played a role, the exact process that led to this cluster of childhood cancers is still unknown.

14.2 Interpret A Correlation Coefficient Using A p-value Or Critical Value

Formulating the hypotheses

You can use a hypothesis test to determine whether the product-moment correlation coefficient r for a particular sample indicates that there is likely to be a linear relationship within the whole population, i.e. whether there is correlation.

As with any hypothesis test, you must define the null and alternative hypotheses.

To test whether the population PMCC ρ is greater than zero use:

$$H_0: \rho = 0; H_1: \rho > 0$$

This is known as a **right-sided one-tailed test**.

To test whether ρ is less than zero use:

$$H_0: \rho = 0; H_1: \rho < 0$$

This is known as a **left-sided one-tailed test**.

To test whether ρ is not equal to zero use:

$H_0: \rho = 0;\ H_1: \rho \neq 0$

This is a **two-tailed test**.

You can determine a critical value for r using the table of critical values in your formula booklet. The critical value, sometimes written as r_{crit}, depends on both the significance level of the test and the sample size.

Worked Example

1. Find the critical value for r, using a significance level of 5% and a sample size of 6.

Find the table 'Critical Values For Correlation Coefficients' in the formula booklet.

Sample Level	Product Moment Coefficient Level				
	0.10	0.05	0.025	0.01	0.005
4	0.8000	0.9000	0.9500	0.9800	0.9900
5	0.6870	0.8054	0.8783	0.9343	0.9587
6	0.6084	0.7293	0.8114	0.8822	0.9172
7	0.5509	0.6694	0.7545	0.8329	0.8745
8	0.5067	0.6215	0.7067	0.7887	0.8343

From the table, the critical value for r is 0.7293.

Note: For a sample size of 6 you see from the table that the critical value of r, with a significance level of 5% for a one-tailed test, is 0.7293. An observed value of r greater than this from a sample of size 6 would provide sufficient evidence to reject the null hypothesis and conclude that $\rho > 0$. Similarly, an observed value of r less than −0.7293 would provide sufficient evidence to conclude that $\rho < 0$.

You reject the null hypothesis if the magnitude of the observed r value is greater than the critical r value.

Worked Example

2. Joel makes bread at home. Every day for 2 weeks he measures the time he lets the dough rise and the final height of the loaf when it comes out of the oven. The results are summarised in the tables below.

Week 1	Day						
	1	2	3	4	5	6	7
Time to rise (mins)	65	80	100	90	55	85	120
Height of loaf (cm)	12	14	16	15	13	14	15

Week 2	Day						
	1	2	3	4	5	6	7
Time to rise (mins)	85	125	65	70	95	55	70
Height of loaf (cm)	13.5	15	12	12	16	12.5	14

(a) Find the product-moment correlation coefficient.

Joel believes there is a positive correlation between the time he gives the loaf to rise and the height of the loaf.

(b) Should Joel carry out a one-tailed or two-tailed test? If one-tailed, is it right-sided or left-sided?

(c) Write down suitable null and alternative hypotheses.

(d) Carry out a hypothesis test to determine whether Joel's belief is correct. Use a significance level of 5%.

(a) From the calculator $r = 0.772$ (3 s.f.)

(b) Joel believes there is a positive correlation between the time he gives a loaf to rise and the height of the loaf. He should carry out a right-sided one-tailed test.

(c) $H_0: \rho = 0;\ H_1: \rho > 0$

(d) There are 14 pairs of data, so $n = 14$

From the table, using $n = 14$ and a significance level of 0.05, the critical r value $r_{crit} = 0.4575$.

$|r| > r_{crit} \therefore$ reject H_0 and accept H_1. There is evidence at the 5% significance level of correlation.

Joel can conclude that there is a significant positive correlation between the time to rise and the height of his loaves.

In the following example, a two-tailed test is carried out. The significance level of 10% is halved, to find a probability of 0.05 in each tail.

Worked Example

3. A traffic researcher takes measurements of air pollution and the number of cars on the road over a 30 day period. Using these 30 pairs of measurements, she obtains a product-moment correlation coefficient of 0.588. The researcher believes there is no correlation between the number of cars on the road and the air pollution. Test the researcher's belief at a 10% level of significance.

Form the null and alternative hypotheses:

$H_0: \rho = 0$

$H_1: \rho \neq 0$

The sample size is 30.

We are using a 10% level of significance, i.e. $\alpha = 0.1$. Since this is a two-tailed test, halve the level of significance to find the probability in each tail. We use a significance level of 0.05 in each tail.

From the table, when $n = 30$ and using a significance level of 0.05, we find a critical r value of 0.3061.

Since $0.588 > 0.3061$, we reject the null hypothesis and accept the alternative hypothesis.

We conclude there is evidence that $\rho \neq 0$, i.e. there is evidence of a correlation between the levels of air pollution and the number of cars on the road.

Exercise 14B

1. In each case determine whether the null hypothesis should be rejected. Show all the steps in your hypothesis test.
   (a) $r = 0.6$; $n = 6$; $\alpha = 0.1$;
      type of test: one-tailed, right-sided
   (b) $r = 0.755$; $n = 20$; $\alpha = 0.01$;
      type of test: two-tailed
   (c) $r = 0.572$; $n = 15$; $\alpha = 0.05$;
      type of test: one-tailed, left-sided
   (d) $r = 0.2$; $n = 100$; $\alpha = 0.1$;
      type of test: two-tailed
   (e) $r = 0.3$; $n = 40$; $\alpha = 0.025$;
      type of test: one-tailed, right-sided

2. In a sandcastle competition, nine teams each build a sandcastle on the beach with a flag on top. As the tide comes in, the sandcastles are slowly washed away. The team whose flag falls last wins the competition.

   When building is complete, Grace's team has the smallest castle. She is worried that there may be a positive correlation between the height of the castle and the time taken for the flag to fall.

   The tide comes in and the castles are washed away. The results are shown in the following table:

Exercise 14B...

Team name	Height of castle (cm)	Time for flag to fall (minutes)
Archie's Aces	104	27
Bella's Builders	113	25
Charlie's Castle Crew	138	26
Donal's Diggers	87	16
Erin's Earthmovers	144	20
Finn's Flag Flyers	113	30
Grace's Grafters	72	21
Harry's Heroes	120	21
Ian's Incredibles	78	23

(a) Calculate the product-moment correlation coefficient for the height and time.
(b) Test Grace's belief that a correlation exists between the height and the time using a significance level of 10%. Show your hypotheses clearly and state your conclusion.

3. A population of drivers takes two different driving tests. A sample of 50 drivers was taken from the population and their scores on the two tests were recorded. A product-moment correlation coefficient of 0.301 was calculated. Test whether or not this shows evidence of correlation between the test scores:
   (a) at the 5% level,
   (b) at the 2% level.

   Hint: 'Evidence of correlation' could mean either positive or negative correlation, so you need to use a two-tailed test with $H_0: \rho = 0$ and $H_1: \rho \neq 0$.

4. The table below shows the number of times a technician checks his work against the number of errors that occur during an eight-week period.

Number of checks, x	6	8	9	2	4	5	6	7
Number of errors, y	5	2	1	10	6	5	5	3

(a) Calculate the product-moment correlation coefficient for these data.

Exercise 14B...

(b) For these data test the null hypothesis that there is no correlation against a suitable alternative hypothesis using a 1% significance level.

5. The ages in months and the weights in kilograms of a random sample of 10 babies are shown in the table below.

Age (months)	2	4	3	2	3	3	4	3	3	2
Weight (kg)	5.2	5.4	6.8	5.0	4.4	6.1	7.6	4.2	6.3	6.8

The product moment correlation coefficient between weight and age for these babies was found to be 0.240 (3 s.f.).

By testing for positive correlation at the 5% significance level interpret this value.

6. The table below shows the daily maximum air temperature and the air pressure during one week in Belfast.

Temp. (°C)	15.5	16.5	16.0	22.6	20.2	21.1	20.3
Pressure (hPa)	1005	1006	1007	992	1004	993	1002

There is found to be a product-moment correlation coefficient of $r = -0.841$ between the air temperature and pressure during this week.

Test, at the 1% level of significance, the claim that there is a negative correlation.

7. A recruitment company carries out a survey of workers in various industries. Each person's age in years and salary is recorded.

Age	29	36	50	31	46	22	19	33	41
Salary (£1000s)	16.5	22.0	35.5	29.0	46.0	15.0	20.0	45.0	31.0

(a) Using the data in the table, find the product-moment correlation coefficient between age and salary.

(b) The recruitment company claims there is no correlation between age and salary. A worker disputes this, claiming there is some correlation. Test the worker's claim at the 5% significance level, stating your null and alternative hypotheses clearly.

Exercise 14B...

8. Twenty teams play in a hockey league. The diagram below shows a scatter graph representing the 'Goals for' and the 'Goals against' for each of these teams during last season.

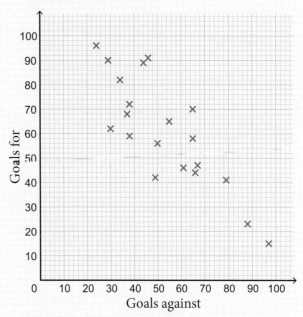

(a) Why would you expect the product-moment correlation coefficient to be negative?

(b) The product-moment correlation coefficient is found to be $r = -0.825$. Using the table of Critical Values for Correlation Coefficients, test this value of r at a 1% significance level.

9. Caitlin is a family doctor. She wishes to investigate whether there is a link between high blood pressure and being overweight. The table below shows the blood pressure and weight of six of her patients.

Systolic blood pressure (mmHg)	135	140	127	132	160	180
Weight (kg)	65	80	62	75	67	85

(a) Calculate the product-moment correlation coefficient between the blood pressures and the weights of these patients.

Caitlin suspects there is a positive correlation between the blood pressure and weight of her patients. She decides to carry out a one-tailed

Exercise 14B...

hypothesis test at the 0.5% level of significance to determine whether there is evidence to support this.

(b) Identify the null and alternative hypotheses Caitlin should use.

(c) Complete the hypothesis test.

(d) Caitlin is surprised at her result in part (c). Explain what change she could make to her test to obtain a different outcome.

10. Steven is a vet. For a research project he studies whether there is a link between a dog's mass and its lifespan. He carries out a hypothesis test and states the following hypotheses:

$H_0: \rho = 0$
$H_1: \rho > 0$

(a) Is this a one-tailed or two-tailed hypothesis test?

(b) For a sample of 15 dogs, Steven calculates the product-moment correlation coefficient to be 0.612. Carry out Steven's hypothesis test at the 5% significance level and state his conclusions.

14.3 Summary

When conducting a hypothesis test for a correlation coefficient, begin by writing down the null and alternative hypotheses. These should always relate to ρ, the population correlation coefficient. As in previous hypothesis tests, H_0 should involve an equality, usually $H_0: \rho = 0$.

The alternative hypothesis depends on the type of test. If it is a one-tailed test, then use either $H_1: \rho > 0$ or $H_1: \rho < 0$

If a two-tailed test is being carried out, then use $H_1: \rho \neq 0$

The table of Critical Values for Correlation Coefficients in your formula booklet is used to find critical r values. These are compared with the value of r observed in a sample or experiment.

If the magnitude of the observed r value is greater than the critical r value, then the observed correlation is significant.

If a two-tailed test is being performed, the significance level should be halved, and this value used to look up the critical r value.

Answers

Exercise 1A

1. $s = ?; u = 0; a = 1; t = 3$

   $s = ut + \dfrac{1}{2}at^2$

   $s = 0(3) + \dfrac{1}{2}(1)(3^2)$

   $s = 4.5$ m. The postman can only run 4.5 metres in 3 seconds, so he doesn't reach the gate in this time.

2. (a) $\dfrac{dy}{dt} = 16t - 2t^{-2}$ (b) $\dfrac{1}{2}$ (c) 20

   (d) $\dfrac{d^2y}{dt^2} = 16 + 4t^{-3}$

   When $t = \dfrac{1}{2}, \dfrac{d^2y}{dt^2} = 48 > 0 \therefore$

   $y = 20$ is a minimum value of y.

3. (a) $|\overrightarrow{PQ}| = \sqrt{7^2 + 1^2} = \sqrt{50}$

   $|\overrightarrow{QR}| = \sqrt{(-5)^2 + 5^2} = \sqrt{50}$

   $|\overrightarrow{RP}| = \sqrt{(-2)^2 + (-6)^2} = \sqrt{40}$

   $|\overrightarrow{PQ}| = |\overrightarrow{QR}| \neq |\overrightarrow{RP}| \therefore$ PQR is isosceles

   (b) 20 cm²

Exercise 1B

1. (a) $v = \dfrac{5}{2}t^2$ (b) $s = \dfrac{5}{6}t^3$

2. (a) 16 m s⁻¹ (b) 12 s

3. $T = \dfrac{2}{3}$

4. (a) $t_1 = \dfrac{1}{3}, t_2 = 2.5$

   (b) $t = \dfrac{17}{12} = 1.42$ s (c) 14.1 m

5. (a) $t = 0, 2$ s (b) 18.25 m (c) 1.08 s

6. 6 seconds

7. 3.23 s

8. (a) 0.3 m s⁻¹ (b) 2 s (c) 0.8 m

9. (a) $v(t) = t(t - 4)(t^2 - 4t + 1)$

   (b) $t = 0, 4, 2 \pm \sqrt{3}$ s

   (c) $a = 4t^3 - 24t^2 + 34t - 4$

   (d) $2 \pm \dfrac{\sqrt{14}}{2}$ or 0.129 s and 3.87 s

   (e) 2 s

10. (a) 6.08 s (b) 33.2 m (c) The rocket is modelled as a particle and/or there is no air resistance.

Exercise 1C

1. (a) $\mathbf{v} = 6t^2\mathbf{i} - 6t^2\mathbf{j}$
   (b) $\mathbf{s} = 2t^3\mathbf{i} - 2t^3\mathbf{j}$

2. (a) $(48\mathbf{i} - 44\mathbf{j})$ m s⁻¹
   (b) (i) $(8\mathbf{i} + 6\mathbf{j})$ m
   (ii) 10 m

3. (a) 10 m s⁻¹ (b) 10 m

4. (a) $\mathbf{a} = 0.2t\mathbf{i} + 0.4t\mathbf{j}$

   $\mathbf{v} = 0.1t^2\mathbf{i} + 0.2t^2\mathbf{j}$

   $\mathbf{r} = \dfrac{1}{30}t^3\mathbf{i} + \dfrac{1}{15}t^3\mathbf{j} + \mathbf{c}$

   When $t = 0, \mathbf{r} = 4\mathbf{j} \therefore \mathbf{c} = 4\mathbf{j}$

   $\mathbf{r} = \dfrac{1}{30}t^3\mathbf{i} + \dfrac{1}{15}t^3\mathbf{j} + 4\mathbf{j}$

   $\mathbf{r} = \dfrac{t^3}{30}\mathbf{i} + \left(\dfrac{t^3}{15} + 4\right)\mathbf{j}$ m

   (b) 5.87 m (c) The player is modelled as a particle and/or there is no air resistance.

5. (a) 4 (b) (i) $\mathbf{a} = (-24t\mathbf{i} + 8\mathbf{j})$ m s⁻²
   (ii) 36.9 or $4\sqrt{85}$ m s⁻²

6. (a) $(30\mathbf{i} - 3\mathbf{j})$ m s⁻²

   (b) $\mathbf{s} = \left(t^3\mathbf{i} - \dfrac{3}{2}t^2\mathbf{j}\right)$ m

   Particle at O when $\mathbf{s} = 0\mathbf{i} + 0\mathbf{j}$

   $t^3\mathbf{i} - \dfrac{3}{2}t^2\mathbf{j} = 0\mathbf{i} + 0\mathbf{j}$

   $\mathbf{i}$ component: $t^3 = 0$

   $\mathbf{j}$ component: $-\dfrac{3}{2}t^2 = 0$

   Only solution is $t = 0$

7. (a) $(6t^2 - 9t + 5)\mathbf{i} + (8t - 1)\mathbf{j}$ m
   (b) (i) 5 seconds (ii) 6.8
   (iii) $(110\mathbf{i} + 39\mathbf{j})$ m

8. (a) (i) $\mathbf{a} = 6(t - 1)\mathbf{j} = (6t - 6)\mathbf{j}$

   $\mathbf{v} = \displaystyle\int \mathbf{a}\,dt = (3t^2 - 6t)\mathbf{j} + \mathbf{c}$

   Initially, P has velocity $2\mathbf{i} + 3\mathbf{j}$.
   $\therefore 2\mathbf{i} + 3\mathbf{j} = 0\mathbf{j} + \mathbf{c}$

   $\mathbf{c} = 2\mathbf{i} + 3\mathbf{j}$

   $\mathbf{v} = (3t^2 - 6t)\mathbf{j} + 2\mathbf{i} + 3\mathbf{j}$

   $= 2\mathbf{i} + (3t^2 - 6t + 3)\mathbf{j}$

   $= 2\mathbf{i} + 3(t^2 - 2t + 1)\mathbf{j}$ m s⁻¹

   (ii) 1 s

   (b) (i) $\mathbf{r} = \displaystyle\int \mathbf{v}\,dt$

   $= \displaystyle\int 2\mathbf{i} + (3t^2 - 6t + 3)\mathbf{j}\,dt$

   $\mathbf{r} = 2t\mathbf{i} + (t^3 - 3t^2 + 3t)\mathbf{j} + \mathbf{c}$

Initially, P has position vector $-\mathbf{i} - 4\mathbf{j}$.
   $\therefore -\mathbf{i} - 4\mathbf{j} = 0\mathbf{i} + 0\mathbf{j} + \mathbf{c}$

   $\mathbf{c} = -\mathbf{i} - 4\mathbf{j}$

   $\mathbf{r} = 2t\mathbf{i} + (t^3 - 3t^2 + 3t)\mathbf{j} + -\mathbf{i} - 4\mathbf{j}$

   $\mathbf{r} = (2t - 1)\mathbf{i} + (t^3 - 3t^2 + 3t - 4)\mathbf{j}$

   (ii) $(x - 1)^3 = 3$

   $(x - 1)(x - 1)^2 = 3$

   $(x - 1)(x^2 - 2x + 1) = 3$

   $x^3 - 3x^2 + 3x - 1 = 3$

   $x^3 - 3x^2 + 3x - 4 = 0$

   (iii) $t = \sqrt[3]{3} + 1 = 2.44$ s (3 s.f.)

9. (a) Distance

   $= |\mathbf{r}| = \sqrt{(2\sin t)^2 + (2\cos t)^2}$

   $= \sqrt{4\sin^2 t + 4\cos^2 t}$

   $= \sqrt{4(\sin^2 t + \cos^2 t)}$

   $= \sqrt{4(1)} = 2$ m

   (b) $\mathbf{r} = 2\mathbf{j}$ m

   (c) $\mathbf{v} = \dfrac{d\mathbf{r}}{dt}$

   $= (2\cos t)\mathbf{i} - (2\sin t)\mathbf{j}$ m s⁻¹

   (d) Speed $= |\mathbf{v}|$

   $= \sqrt{(2\cos t)^2 + (-2\sin t)^2}$

   $= \sqrt{4\cos^2 t + 4\sin^2 t}$

   $= \sqrt{4(\cos^2 t + \sin^2 t)}$

   $= \sqrt{4(1)} = 2$ m s⁻¹

10. $\dfrac{1}{4}(\mathbf{i} - \mathbf{j})$ m

11. (a) When $t = 0$: $\mathbf{r} = 3(3 - 3)^2\mathbf{i} + (3(3 - 3) - (3 - 3)^3)\mathbf{j}$
    $\mathbf{r} = 0\mathbf{i} + 0\mathbf{j}$ m, which is the position vector of point O.
    (b) $3(t - 3) - (t - 3)^3 = 0$
    $(t - 3)(3 - (t - 3)^2) = 0$
    $t = 3$ or $3 - (t - 3)^2 = 0$
    $t = 3$ relates to the particle passing through O. So at A,
    $3 - (t - 3)^2 = 0$
    $(t - 3) = \pm\sqrt{3}; t = 3 \pm \sqrt{3}$
    (c) $9\mathbf{i}$ m
    (d) $\mathbf{v} = 6(t - 3)\mathbf{i} + (3 - 3(t - 3)^2)\mathbf{j}$
    Speed $= = |\mathbf{v}|$
    $= \sqrt{36(t - 3)^2 + (3 - 3(t - 3)^2)^2}$
    $= \sqrt{36(t - 3)^2 + 9 - 18(t - 3)^2 + 9(t - 3)^4}$

$$= \sqrt{9(t-3)^4 + 18(t-3)^2 + 9}$$
$$= 3\sqrt{(t-3)^4 + 2(t-3)^2 + 1}$$
$$= 3\sqrt{((t-3)^2 + 1)^2}$$
$$= 3((t-3)^2 + 1)$$
$$= 3(t^2 - 6t + 10)$$

Exercise 2A

1. 45 m s^{-1}
2. (a) $(7\mathbf{i} + 9.5\mathbf{j}) \text{ m}$
   (b) 11.8 m (3 s.f.) (c) 5.41 m s^{-1}

Exercise 2B

1. 1.88 m
2. In the horizontal:
   $$\text{time} = \frac{\text{distance}}{\text{velocity}} = \frac{12}{20} = 0.6$$
   In the vertical:
   $s = ?; u = 0; a = g; t = 0.6$
   $$s = ut + \frac{1}{2}at^2$$
   $$= 0(0.6) + \frac{1}{2}(9.8)(0.6^2)$$
   $$= 1.764 \text{ m}$$
   The ball begins at a height of 2.5 m and travels 1.764 m vertically downwards.
   2.5 – 1.764 = 0.736 m
   By the time the ball has travelled 12 m horizontally, it is at a height of 0.736 m, so it hits the net.
3. 0.975 m
4. (a) 0.306 cm (b) 20.0 m
5. (a) $^{15}/_7$ s or 2.14 s (b) 15 m
   (c) 1. The ground is horizontal.
   2. The football is modelled as a particle and/or there is no air resistance.
6. 42 m s^{-1}
7. (a) 27 m (b) 27 m
8. (a) For particle projected from A, in vertical direction:
   $s = 2.5; u = 0; a = 9.8; t = ?$
   $$s = ut + \frac{1}{2}at^2 \Rightarrow 2.5 = 4.9t^2$$
   $$\Rightarrow t = \frac{5}{7} \text{ s}$$
   For particle projected from B, in vertical direction:
   $s = 14.4; u = 0; a = 9.8; t = ?$
   $$s = ut + \frac{1}{2}at^2 \Rightarrow 14.4 = 4.9t^2$$

$$\Rightarrow t = \frac{12}{7} \text{ s}$$

Particle from B takes 1 s longer, but is released 1 s earlier, so the particles hit ground at the same time.

(b) For particle projected from A, in horizontal direction:
Distance = velocity × time
$$= 12 \times \frac{5}{7} = \frac{60}{7} = 8.57 \text{ m (3 s.f.)}$$
For particle projected from B, in horizontal direction:
Distance = velocity × time
$$= 5 \times \frac{12}{7} = \frac{60}{7} = 8.57 \text{ m (3.s.f)}$$
Both particles land 8.57 m from the tower.

9. (a) Let t_1 and t_2 be the times of flight for first and second stones respectively.
   Vertically for 1st stone:
   $s = 9h; u = 0; a = 9.8; t = t_1$
   $$s = ut + \frac{1}{2}at^2 \Rightarrow 9h = 4.9t_1^2 \quad (1)$$
   Vertically for 2nd stone:
   $s = 4h; u = 0; a = 9.8; t = t_2$
   $$s = ut + \frac{1}{2}at^2 \Rightarrow 4h = 4.9t_2^2 \quad (2)$$
   $$(1) \div (2) \Rightarrow \frac{9}{4} = \frac{t_1^2}{t_2^2} \Rightarrow \frac{t_1}{t_2} = \frac{3}{2} \quad (3)$$
   Horizontally for 1st stone:
   distance = speed × time
   $$\Rightarrow 0.8AB = U_1 t_1 \quad (4)$$
   Horizontally for 2nd stone:
   distance = speed × time
   $$\Rightarrow 0.2AB = U_2 t_2 \quad (5)$$
   Eliminating AB from (4) and (5) gives:
   $$U_1 t_1 = 4U_2 t_2 \Rightarrow \frac{t_1}{t_2} = \frac{4U_2}{U_1} \quad (6)$$
   (3) and (6)
   $$\Rightarrow \frac{4U_2}{U_1} = \frac{3}{2} \Rightarrow 3U_1 = 8U_2$$
   (b) Both stones are modelled as particles.

Exercise 2C

1. (a) 1.3 m (b) 14.8 m
2. Range $= \dfrac{u^2 \sin 2\theta}{g}$

$$R = \frac{u^2 \sin(2 \times 15°)}{g}$$
$$= \frac{u^2 \sin 30°}{g} = \frac{u^2 \times \frac{1}{2}}{g}$$
$$\Rightarrow gR = \frac{1}{2}u^2 \Rightarrow u^2 = 2gR$$

3. (a) 45° (b) The maximum range occurs when the launch angle $\theta = 45°$. Using the equation of the trajectory, find the initial speed u required for a rock to reach the town, where $x = 2000$ and $y = -500$, using $\theta = 45°$.
   $$-500 = 2000 \tan 45° - \frac{9.8 \times 2000^2}{2u^2 \cos^2 45°}$$
   $$-500 = 2000(1) - \frac{39\,200\,000}{2u^2(0.5)}$$
   $$-2500 = -\frac{39\,200\,000}{u^2}$$
   $$u^2 = \frac{39\,200\,000}{2500}$$
   $$u^2 = 15680$$
   $$u = 125 \text{ m s}^{-1} \text{ (3 s.f.)}$$
   The launch speed required for a rock to reach the town, given a launch angle of 45°, is 125 m s^{-1}, so rocks from the volcano will not directly hit the town.

4. $\cos\theta = \dfrac{12}{13}$ and $\sin\theta = \dfrac{5}{13}$
   The horizontal component of the velocity is
   $$u\cos\theta = 30 \times \frac{12}{13} = \frac{360}{13} \text{ m s}^{-1}.$$
   The vertical component of the velocity is
   $$u\sin\theta = 30 \times \frac{5}{13} = \frac{150}{13} \text{ m s}^{-1}.$$
   In the horizontal,
   $$\text{velocity} = \frac{\text{distance}}{\text{time}} \Rightarrow \frac{360}{13} = \frac{16}{t}$$
   $$\Rightarrow t = \frac{26}{45} \text{ s}$$
   In the vertical,
   $$s = ?; u = \frac{150}{13}; a = -9.8; t = \frac{26}{45}$$
   $$s = ut + \frac{1}{2}at^2$$
   $$s = \frac{150}{13} \times \frac{26}{45} + \frac{1}{2}(-9.8)\left(\frac{26}{45}\right)^2$$

= 5.03 m (3 s.f.)
When the golf ball has travelled 16 metres horizontally, it has a vertical height of 5.03 m. Since the tree is 3 m tall, the golf ball clears the tree by 2.03 m.

5. (a) $25\sqrt{3}$ m (b) 12 m

6. (a) 32.8° (b) 55.8 m (c) 27.4 m s⁻¹

7. $2\sqrt{2g}$

8. $\dfrac{289}{2g}$

9. (a) Consider the vertical motion, taking downwards as the positive direction. The initial vertical velocity is $u \sin\theta$.
$s = y$; $u = u\sin\theta$; $a = g$; $t = t$

$s = ut + \dfrac{1}{2}at^2$

$\Rightarrow y = ut\sin\theta + \dfrac{1}{2}gt^2$ (1)

Consider the horizontal motion. The horizontal component of the velocity is $u\cos\theta$, and this is constant throughout the motion. For the horizontal motion with constant velocity,

$\text{velocity} = \dfrac{\text{distance}}{\text{time}}$

$\Rightarrow u\cos\theta = \dfrac{x}{t}$

$\Rightarrow t = \dfrac{x}{u\cos\theta}$ (2)

Substitute into (1):

$y = u\left(\dfrac{x}{u\cos\theta}\right)\sin\theta + \dfrac{1}{2}g\left(\dfrac{x}{u\cos\theta}\right)^2$

$\Rightarrow y = \dfrac{ux\sin\theta}{u\cos\theta} + \dfrac{1}{2}g\left(\dfrac{x^2}{u^2\cos^2\theta}\right)$

$\Rightarrow y = x\tan\theta + \dfrac{gx^2}{2u^2\cos^2\theta}$

$y = x\tan\theta + \dfrac{gx^2}{2u^2}\sec^2\theta$

(b) If $x = y$, then

$x = x\tan\theta + \dfrac{gx^2}{2u^2}\sec^2\theta$

$1 = \tan\theta + \dfrac{gx}{2u^2}\sec^2\theta$

$2u^2 = 2u^2\tan\theta + gx\sec^2\theta$

$2u^2 - 2u^2\tan\theta = gx\sec^2\theta$

$x = \dfrac{2u^2(1-\tan\theta)}{g\sec^2\theta}$

$x = \dfrac{2u^2}{g}\cos^2\theta\,(1-\tan\theta)$

From (2), $t = \dfrac{x}{u\cos\theta}$

$\therefore t = \dfrac{2u}{g}\cos\theta\,(1-\tan\theta)$

or $t = \dfrac{2u}{g}(\cos\theta - \sin\theta)$

10. (a) Vertically: $v^2 = u^2 + 2as$
$(13\sin 39)^2 = (u\sin\theta)^2 - 2(9.8)(3)$
$\Rightarrow u\sin\theta = \sqrt{125.73\ldots} = 11.21\ldots$
Horizontally:
$u\cos\theta = 13\cos 39 = 10.10\ldots$

$\dfrac{u\sin\theta}{u\cos\theta} = \dfrac{11.21\ldots}{10.10\ldots}$

$\tan\theta = 1.10988\ldots$
$\theta = 47.98\ldots = 48.0°$ (3 s.f.)
Since $u\cos\theta = 10.10\ldots$

$u = \dfrac{10.10\ldots}{\cos(47.98\ldots)} = 15.09\ldots$

$= 15.1$ m s⁻¹ (3 s.f.)

(b) Vertically: $s = ut + \dfrac{1}{2}at^2$

$3 = (11.21\ldots)t - \dfrac{1}{2}(9.8)t^2$

$4.9t^2 - (11.21\ldots)t + 3 = 0$
$t = 0.309, 1.978$
$t = 1.98$ s (3 s.f.) when ball returns to a height of 3 m
(c) The rugby ball is modelled as a particle; or air resistance is ignored.

11. Peter's statement is incorrect: it is possible for the stone to bounce back and hit him or Lucy. It is true that the stone loses energy as it strikes the rock, which means that the speed with which the stone bounces back will be lower than the speed with which it struck the rock. However, the distance travelled depends on both the speed and the angle at which it bounces off the rock. If the angle of its return path is 45° to the horizontal, the range of the stone is maximised. It is possible for the range to be greater than the distance between the thrower and the rock despite a lower speed.

Exercise 2D

1. $b = 15$, $t = 1$

2. (a) 0.507 s (b) The ball is modelled as a particle and/or air resistance is ignored.

3. (a) 3 s (b) 15 m (c) 30 m s⁻¹ (2 s.f.)
(d) The effect of folding the wings is to reduce air resistance to a minimum.

4. (a) Consider the book's journey from the teacher. We will work out its vertical position at the time it has travelled 4 m in the horizontal direction.
$\mathbf{s} = 4\mathbf{i} + x\mathbf{j}$; $\mathbf{u} = 7\mathbf{i} + 2.1\mathbf{j}$; $\mathbf{v} = \ $;
$\mathbf{a} = -9.8\mathbf{j}$; $t = T$

$\mathbf{s} = \mathbf{u}t + \dfrac{1}{2}\mathbf{a}t^2$

$4\mathbf{i} + x\mathbf{j} = (7\mathbf{i} + 2.1\mathbf{j})T + \dfrac{1}{2}(-9.8\mathbf{j})T^2$

i component: $4 = 7T \Rightarrow T = \dfrac{4}{7}$

(b) It takes ⁴⁄₇ s for the book to travel 4 m horizontally.
j component: $x = 2.1T - 4.9T^2$

$= 2.1\left(\dfrac{4}{7}\right) - 4.9\left(\dfrac{4}{7}\right)^2 = -0.4$

When the book has a horizontal displacement of 4 m, it has a vertical displacement of –0.4 m, i.e. its displacement from the teacher is $(4\mathbf{i} - 0.4\mathbf{j})$ m. It lands on Jack's desk.
(c) The textbook is modelled as a particle and/or the effects of air resistance are ignored. Air resistance slows the book down as it travels, so it may not maintain its horizontal velocity of 4 m s⁻¹. Therefore, it may fall under the desk.

5. (a) $\mathbf{s} = 6\mathbf{i} + x\mathbf{j}$; $\mathbf{u} = 15\mathbf{i} + 0.5\mathbf{j}$; $\mathbf{a} = -9.8\mathbf{j}$; $t = T$

$\mathbf{s} = \mathbf{u}t + \dfrac{1}{2}\mathbf{a}t^2$; $6\mathbf{i} + x\mathbf{j}$

$= (15\mathbf{i} + 0.5\mathbf{j})T + \dfrac{1}{2}(-9.8\mathbf{j})T^2$

i component: $6 = 15T \Rightarrow T = 0.4$
It takes 0.4 seconds for the pellet to hit Monty.
(b) j component:
$x = 0.5T - 4.9T^2 = -0.584$

The pellet's displacement from its starting position is $6\mathbf{i} - 0.584\mathbf{j}$. Its height is $1.25 - 0.584 = 0.666$ m above the ground, i.e. on Monty's legs.

Exercise 3A

1. (a) (i) A particle is a theoretical object that has a mass but no size (it is zero dimensional). (ii) A rod is a theoretical one-dimensional object; that is, it has a length, but no width or height. (b) A light rod is a rod whose mass is negligible.
2. (a) $a = 13.0$ cm; $b = 12.0$ cm
   (b) $c = 8.02$ cm; $d = 4.09$ cm
3.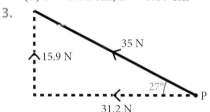
4. (a)
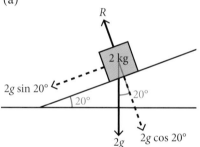

(b) Parallel component:
$2g \sin 20° = 6.70$ N
Perpendicular component:
$2g \cos 20° = 18.4$ N

Exercise 3B

1. (a) 10 N m anticlockwise
   (b) 12 N m clockwise
   (c) 20.5 N m clockwise
   (d) 70 N m anticlockwise
2. (a) Taking moments about point O: Clockwise moment from 6 N force: $6 \times 3 = 18$ N m
   Anticlockwise moment from 2 N force: $2 \times 1 = 2$ N m
   Resultant moment: 16 N m clockwise.
   In addition, resolving forces perpendicular to the rod, there is a resultant force of $2 + 6 - 5$ N acting. The rod is not in

equilibrium.
(b) Resolving forces perpendicular to the rod: there is no resultant force. However, taking moments about point C, the 60 N force provides an anticlockwise moment of $60 \times 4 = 240$ N m. The other two forces provide no moment. Therefore the rod is not in equilibrium.
(c) Resolving forces perpendicular to the rod: $250 = 150 + 100$, therefore there is no resultant force. Taking moments about point Q: the 150 N force provides a clockwise moment of $150 \times 2 = 300$ N m. The 100 N force provides an anticlockwise moment of $100 \times 3 = 300$ N m. Taking moments about any other point again shows there is no resultant moment. Therefore, the rod is in equilibrium.

3. (a) $F_1 = 20$ N; $F_2 = 40$ N
   (b) $F_1 = 100$ N; $F_2 = 80$ N
   (c) $F_1 = 8$ N; $F_2 = 12$ N
   (d) $F_1 = 5$ N; $F_2 = 24$ N

Exercise 3C

1. (a) Moments about A: Clockwise: $6 \times 2 \sin 135 = 6\sqrt{2}$ N m
   Anticlockwise:
   $\dfrac{3\sqrt{2}}{2} \times 4 = 6\sqrt{2}$ N m
   Also, resolving forces perpendicular to rod:
   $6 \times \cos 45 = 3 \cos 45 + \dfrac{3\sqrt{2}}{2}$
   So the rod is in equilibrium.
   (b) Moments about A: Clockwise: $6 \times 2 \sin 60 + 2 \times 2 = 14.4$ N m (3 s.f.). Anticlockwise: $2 \times 8 = 16$ N m
   Rod is not in equilibrium.
2. (a) 5 N m clockwise
   (b) $4\sqrt{3} - 1 = 5.93$ N m clockwise
   (c) 10 N m clockwise
   (d) 0 N m (the rod is in equilibrium)

(e) 5.71 N m anticlockwise
3. (a) $F_1 = \dfrac{6 + 4\sqrt{3}}{3} = 4.31$ N (3 s.f.);
   $F_2 = \dfrac{20 - 2\sqrt{3}}{3} = 5.51$ N (3 s.f.)
   (b) $F_1 = \dfrac{4\sqrt{3}}{3} = 2.31$ N (3 s.f.);
   $F_2 = \dfrac{2\sqrt{3}}{3} = 1.15$ N (3 s.f.)

Exercise 4A

1. There is no resultant force and no resultant moment.
2. 35 N m clockwise
3. $F_1 = \dfrac{15\sqrt{3}}{2}$ N or 13.0 N (3 s.f);
   $F_2 = \dfrac{35\sqrt{3}}{2}$ N or 30.3 N (3 s.f)

Exercise 4B

1. $R_A = \dfrac{g}{4}$ N; $R_B = \dfrac{3g}{4}$ N
2. (a)
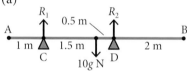
(b) $R_1 = 2.5g$ N; $R_2 = 7.5g$ N
3. (a)
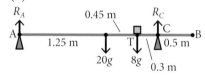
(b) $R_C = 19.3g = 189$ N (3 s.f.);
$R_A = 8.7g = 85.3$ N (3 s.f.)
(c) The plank is modelled as a rod. The toolbox is modelled as a particle. The supports are assumed to be sharp.
4. (a)

R_B R_D 0.5 m

A   B   C   D   F
1 m   1 m   1.5 m
0.5 m   57g   50g   0.5 m   75g

(b) $R_B = 588$ N;
$R_D = 1200$ N (3 s.f.)
(c) The workmen are both modelled as particles. The supports are assumed to be sharp.
5. (a) $R_C = 60g$; $R_D = 30g$
(b) The beam is modelled as a rod. The gymnast is modelled

as a particle. The supports are assumed to be sharp.

Exercise 4C

1. 1.5 m
2. The tortoise must walk 0.6 m, which takes it 12 seconds.
3. (a)

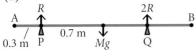

(b) 1.35 m (c) The plank is modelled as a rod. The supports are assumed to be sharp.
4. (a) 0.0392 N and 0.0588 N
   (b) 9.5 cm (c) The pencil is modelled as a rod. The sharpeners are modelled as sharp supports.
5. (a) 41 cm (b) The bench is modelled as a uniform rod and the three people as particles. The legs are assumed to be sharp supports.

Exercise 4D

1. 25 kg
2. (a)

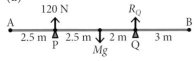

(b) (i) $M = \dfrac{270}{g} = 27.6$ kg (3 s.f.)
(ii) $R_Q = 150$ N
(c) The plank is modelled as a rod. The supports are assumed to be sharp.
3. (a) $M = \dfrac{3R}{g}$ (b) 0.2 m (c) $m = \dfrac{M}{4}$
4. (a) $R_2 = 1000g$ N
   (b) $M = 1500$ kg
   (c) (i) The car is modelled as a particle. (ii) The roadway is modelled as a rod. (iii) The concrete pillars are modelled as sharp supports.
5. (a) 2000
   (b) (i) 75 000 kg (ii)

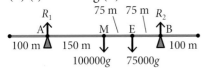

(iii) $R_1 = 673\,750$ N;
$R_2 = 1\,041\,250$ N; or (to 3 s.f.):
$R_1 = 674\,000$ N;
$R_2 = 1\,040\,000$ N
The bridge is safe.
(c) (i) The traffic jam around support B could be modelled as a single particle at point B.
(ii)

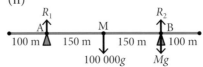

(iii) $R_1 = 490\,000$ N
(iv) If M exceeds roughly 1.02×10^7 kg, the reaction at support B exceeds 10^8 N and the bridge becomes unsafe. This is not likely to happen. Assuming a mean vehicle mass of 1500 kg, this is equivalent to roughly 6800 vehicles in the traffic jam around point B. The maximum number of vehicles that could occupy the entire bridge is roughly 2000, so it is not possible for a traffic jam around point B to comprise 6800 vehicles.

Exercise 4E

1. 33.3 cm (3 s.f.)
2. Tension in string at C: 60.3 N; Tension in string at D: 37.7 N (both to 3 s.f.)
3. (a)

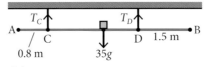

(b) $T_C = 127$ N;
$T_D = 216$ N (3 s.f.)
(c) 0.233 m (3 s.f.)
4. (a)

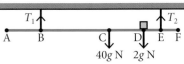

(b) Taking moments about A gives: $(Mg \times x) + (4Mg \times 4x) = (17g \times 5x)$ leading to $M = 5$.
(c) $8g$ N
5. (a) (i) $\dfrac{4}{3}g$ N or 13.1 N (3 s.f.)

(ii) $\dfrac{2}{3}g$ N or 6.53 N
(b) $T = \dfrac{2g}{3}(5y + 2)$
(c) $T > 90 \Rightarrow \dfrac{2g}{3}(5y + 2) > 90$
$5y + 2 > \dfrac{135}{g}$
So without rope breaking
$y \le \dfrac{135 - 2g}{5g}$ m
or $y \le 2.36$ m (3 s.f.)

Exercise 4F

1. 33.3 cm (3 s.f.)
2. $R_P = 9g$; $R_Q = 3g$
3. (a) 2 m
(b)

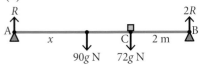

(c) 241 N (3 s.f.) (d) 171 N (3 s.f.)
4. (a)

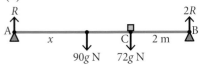

(b) Reaction at A: $54g$ N or 529 N (3 s.f.); Reaction at B: $108g$ N or 1060 N (3 s.f.) (c) 4 m
(d) Any one of: The plank is modelled as a rod. The crate is modelled as a particle. The supports are assumed to be sharp.
5. Resolving vertically:
$R_A + R_B = (M + m)g$ (1)
Moments about A:
$MgpL + mgqL = R_B L$ (2)
$\Rightarrow R_B = (Mp + mq)g$
Sub in (1) for R_B:
$R_A + (Mp + mq)g = (M + m)g$
$R_A = (M + m)g - (Mp + mq)g$
$= (M + m - Mp - mq)g$
$= \big((1 - p)M + (1 - q)m\big)g$

Exercise 4G

1. (a) 0 N (b) 1.5 m (b) Statement i) is correct. A lighter parcel is not heavy enough to make the beam tilt. (Whereas a heavier parcel would cause it to tilt with C as the pivot.)

2. 1:3

3. $\dfrac{8d}{3}$

4. (a)

A $\overset{R_1}{\uparrow}$ ... $\overset{R_2}{\uparrow}$ ... B
2 m $\overset{\blacktriangle}{C}$ 3 m $\underset{80g \text{ N}}{\downarrow}$ 2 m $\overset{\blacktriangle}{D}$ 3 m

(b) $R_1 = 32g$ N; $R_2 = 48g$ N
$R_1 = 314$ N; $R_2 = 470$ N (3 s.f.)

(c)

A ... C ... $\overset{R}{\uparrow}$ ... B□
2 m $\overset{\blacktriangle}{}$ 3 m $\underset{80g \text{ N}}{\downarrow}$ 2 m $\overset{\blacktriangle}{D}$ 3 m $\underset{Mg}{\downarrow}$

(d) $M = \dfrac{160}{3}$ kg or 53.3 kg (3 s.f.)

(e) Any two of: The object is modelled as a particle; the beam is modelled as a rod; the supports are assumed to be sharp.

5. (a) Whole rod length AB is
$d_1 + d_2 + d_3$
Since E is the midpoint of rod,
$EB = \dfrac{1}{2}(d_1 + d_2 + d_3)$
$x = EB - DB$
$= \dfrac{1}{2}(d_1 + d_2 + d_3) - d_3$
$= \dfrac{1}{2}(d_1 + d_2 - d_3)$

(b) $R_1 = \dfrac{Mg}{2d_2}(d_1 + d_2 - d_3)$;

$R_2 = \dfrac{Mg}{2d_2}(d_2 + d_3 - d_1)$

(c) (i) 0 N

(ii) $m = \dfrac{M(d_1 + d_2 - d_3)}{2d_3}$

Exercise 4H

1. 0.8 m

2. (a) 3.25d m (b) With the parcel modelled as a particle, it is modelled as having a mass, but no size. Hence its weight force is concentrated at a single point on the beam.

3. $\dfrac{1}{2g}$ m or 5.1 cm to the right

4. (a) Let additional mass be m kg. When mass at A, moments about X: $mgd_1 = Mg(x - d_1)$ (1)
When mass at B, moments about

Y: $mgd_2 = Mg(l - x - d_2)$ (2)

$(1) \div (2) \Rightarrow \dfrac{d_1}{d_2} = \dfrac{(x - d_1)}{(l - x - d_2)}$

$\Rightarrow d_1(l - x - d_2) = d_2(x - d_1)$

$d_1 l - d_1 x - d_1 d_2 = d_2 x - d_1 d_2$

$d_1 l - d_1 x = d_2 x$

$x(d_1 + d_2) = d_1 l$

$x = \dfrac{d_1 l}{d_1 + d_2}$

(b) $x = \dfrac{d_1 l}{d_1 + d_2}$

If $d_1 = d_2$ then $x = \dfrac{d_1 l}{d_1 + d_1} = \dfrac{l}{2}$.

In this case the centre of mass lies exactly halfway along the beam and therefore the beam is uniform.

5. (a) 79.3 kg (b) Declan's weight shows he is probably an adult. Alice, Ben and Caoimhe could be his children.

Exercise 5A

1. 3.92 N anticlockwise

2. (a)

0.5 m $\overset{R_C}{\uparrow}$ ... M ... $\overset{R_D}{\uparrow}$ 20 kg
A• 1 m $\underset{C}{}$ $\underset{8g}{\downarrow}$ 1 m $\underset{D}{}$ x $\underset{20g}{}$ •B

(b) (i) $R_C = 0$ N; $R_D = 28g$ N or 274 N (3 s.f.) (ii) 0.4 m
(c) The beam is modelled as a rod; the box is modelled as a particle; the supports are sharp.

3. $\mu = \dfrac{5F - 3}{150g}$

Exercise 5B

1. For the force of 25g N,
$25g \times 5 \times \cos 60 = 62.5g$ or 613 N (3 s.f.) clockwise
For the force of 70 N,
$70 \times 10 \times \sin 60 = 606$ N (3 s.f.) anticlockwise

2. 15.6 N

3. (a)

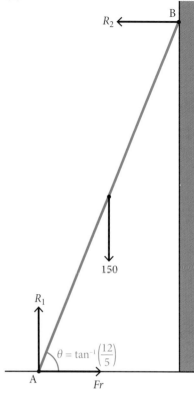

(b) 150 N (c) 31.25 N (d) 31.25 N

4. (a)

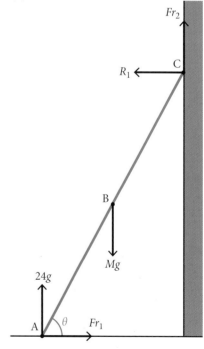

(b) (i) 8g N (ii) 8g N (iii) 2g N
(c) (i) 26 kg (ii) 54.0° (3 s.f.)

5. (a) $2\sqrt{2}g$ N (b) The ladder is modelled as a rod; the string is light and inextensible.

6. (a) 3.07 m (3 s.f.)
(b) 515 N (3 s.f.)

7. (a) The ladder has been labelled AB. Paul is standing at P and the centre of mass is at C.

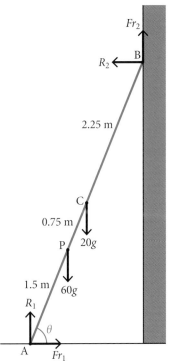

(b) $Fr_2 = 80g - R_1$

(c) Moments about A:
$(60g \times 1.5 \cos \theta) + (20g \times 2.25 \cos \theta)$
$= 4.5R_2 \sin \theta + 4.5Fr_2 \cos \theta$

$90g \cos \theta + 45g \cos \theta$
$\quad = 4.5R_2 \sin \theta + 4.5Fr_2 \cos \theta$

Grouping terms on the LHS and dividing by $\cos \theta$:
$135g = 4.5R_2 \tan \theta + 4.5Fr_2$

Since $\tan \theta = \dfrac{8}{3}$:

$135g = 12R_2 + 4.5Fr_2$

Substitute for Fr_2:
$135g = 12R_2 + 4.5(80g - R_1)$
Expand brackets:
$135g = 12R_2 + 360g - 4.5R_1$
Group terms in g:
$225g = 4.5R_1 - 12R_2$
Divide by 1.5:
$150g = 3R_1 - 8R_2$

(d) $R_1 = 150g$ N

$R_2 = \dfrac{R_1}{4} = 37.5$ N

(e) The ladder is modelled as a rod; Paul is modelled as a particle.

Exercise 5C

1. (a) $20g$ N or 196 N (3 s.f.)
   (b) 34.0 N (3 s.f.)

2. (a)

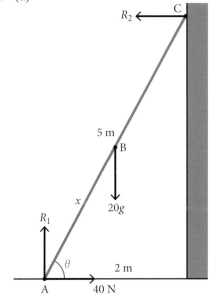

   (b) 2.34 m

3. (a) Moments about foot of ladder:
   $Mg \times 0.9l \times \cos \theta = 6g \times l \times \sin \theta$
   $0.9Mgl \cos \theta = 6gl \sin \theta$
   Divide by $gl \cos \theta$:
   $0.9M = 6 \tan \theta; M = \dfrac{20}{3} \tan \theta$
   (b) 12 kg

4. $\theta = \tan^{-1}\left(\dfrac{8}{7}\right)$

5. (a)

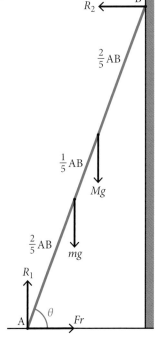

(b) Let the normal reaction forces at the foot and top of the ladder be R_1 and R_2 respectively.
Resolving vertically:
$mg + Mg = R_1$ (1)
Resolving horizontally:
$\dfrac{1}{5}R_1 = R_2$ (2)

Moments about A:
$\dfrac{2}{5} mg \cos \theta + \dfrac{3}{5} Mg \cos \theta = R_2 \sin \theta$

Substituting for R_2 from (2) and multiplying by 5:
$2mg \cos \theta + 3Mg \cos \theta = R_1 \sin \theta$

Substituting for R_1 from (1):
$2mg \cos \theta + 3Mg \cos \theta = (mg + Mg) \sin \theta$

Dividing by $\cos \theta$:
$2mg + 3Mg = (mg + Mg) \tan \theta$

$\tan \theta = \dfrac{2m + 3M}{m + M}$

(c) 69.4° (d) The man is modelled as a particle; the ladder is modelled as a non-uniform rod.

6. (a)

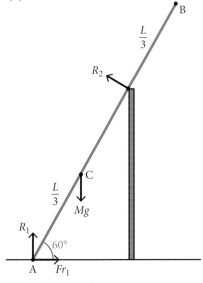

(b) Moments about A:
$R_2 \times \dfrac{2L}{3} = Mg \times \dfrac{L}{3} \times \cos 60$

$\dfrac{2LR_2}{3} = \dfrac{MgL}{3} \times \dfrac{1}{2}$

$4LR_2 = MgL$

$R_2 = \dfrac{Mg}{4}$

(c) $\dfrac{Mg\sqrt{3}}{8}$

Exercise 5D

1. 64.8°

2. (a)

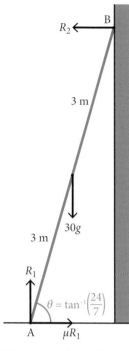

(b) Moments about A:
$$30g \times 3 \times \cos\theta = R_2 \times 6 \times \sin\theta$$
$$\Rightarrow 90g \cos\theta = 6R_2 \sin\theta$$
$$\Rightarrow R_2 = \frac{15g}{\tan\theta} = \frac{15g}{24/7} = \frac{35}{8}g \text{ N}$$

(c) $\dfrac{7}{48}$

3.

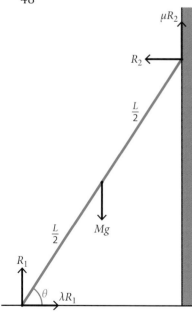

Let M be the mass of the ladder, L be its length. Let R_1 and R_2 be the normal reaction forces at the foot and top of the ladder respectively.
Resolving horizontally:

$R_2 = \lambda R_1 \Rightarrow R_1 = \dfrac{R_2}{\lambda}$ (1)

Resolving vertically:
$R_1 + \mu R_2 = Mg$
Substitute for R_1 from (1):
$$\frac{R_2}{\lambda} + \mu R_2 = Mg \quad (2)$$
Moments about foot of ladder:
$$Mg \times \frac{L}{2} \times \cos\theta = (R_2 \times L \times \sin\theta)$$
$$+ (\mu R_2 \times L \times \cos\theta)$$
Substitute for Mg from (2) and divide by L:
$$\frac{1}{2}\left(\frac{R_2}{\lambda} + \mu R_2\right)\cos\theta = R_2 \sin\theta$$
$$+ \mu R_2 \cos\theta$$
Expanding brackets and divide by R_2:
$$\left(\frac{1}{2\lambda} + \frac{\mu}{2}\right)\cos\theta = \sin\theta + \mu\cos\theta$$
Divide by $\cos\theta$:
$$\frac{1}{2\lambda} + \frac{\mu}{2} = \tan\theta + \mu$$
$$\tan\theta = \frac{1}{2\lambda} - \frac{\mu}{2} = \frac{1 - \mu\lambda}{2\lambda}$$

4. (a) $R_1 = 30g = 294 \text{ N}$ (3 s.f.);
$F = R_2 = 5\sqrt{3}g = 84.9 \text{ N}$ (3 s.f.)

(b) $\dfrac{\sqrt{3}}{6} = 0.289$ (3 s.f.)

5. (a) Let the normal reaction forces at the foot and top of the ladder be R_1 and R_2 respectively.
Resolving vertically:
$mg + Mg = R_1$ (1)
Resolving horizontally:
$\mu R_1 = R_2$ (2)
Moments about A:
$pmg\cos\theta + qMg\cos\theta = R_2\sin\theta$
Substituting for R_2 from (2):
$pmg\cos\theta + qMg\cos\theta = \mu R_1\sin\theta$
Substituting for R_1 from (1):
$pmg\cos\theta + qMg\cos\theta = \mu(mg + Mg)\sin\theta$
Dividing by $\cos\theta$:
$pmg + qMg = \mu(mg + Mg)\tan\theta$
$$\tan\theta = \frac{pm + qM}{\mu(m + M)}$$

(b) $\dfrac{9}{2}$

6. (a)

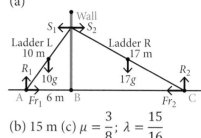

(b) 15 m (c) $\mu = \dfrac{3}{8}$; $\lambda = \dfrac{15}{16}$

Exercise 5E

1. (a)

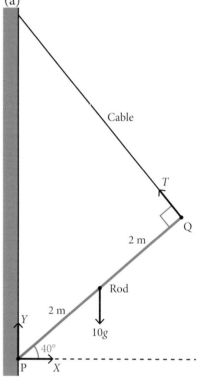

(b) 37.5 N (3 s.f.)

2. (a) $F = 5\sqrt{2}g = 69.3 \text{ N}$ (3 s.f.)

(b) $R = 5\sqrt{2}g = 69.3 \text{ N}$ (3 s.f.);
Direction: $\tan^{-1}(1) = 45°$ to horizontal

3. (a)

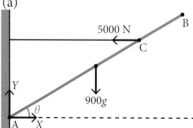

(b) Since the drawbridge is rising, anticlockwise moments are greater than clockwise moments. Let the length of the drawbridge be l metres.

Moments about A:

$$5000 \left(\frac{3}{4} l \right) \sin \theta$$

$$> 900g \left(\frac{l}{2} \right) \sin(90 - \theta)$$

$$3750l \sin \theta > 450lg \cos \theta$$

$$\tan \theta > \frac{450lg}{3750l}$$

$$\tan \theta > \frac{3g}{25}$$

(c) The cable is light; the drawbridge is modelled as a rod; the hinge at point A is smooth.

4. (a) The size of the moment is $Fd \sin \theta$, where F is the size of the force he uses, d is the distance AB and θ is the angle at which he applies the force. An angle of 90° to the bed gives the biggest value of $Fd \sin \theta$, since $\sin \theta$ reaches its maximum value of 1 when $\theta = 90°$. Therefore, Ben should push at B perpendicular to the bed to give the greatest moment.
(b) The bed is modelled as a rod; the hinge is assumed to be smooth.

5. (a)

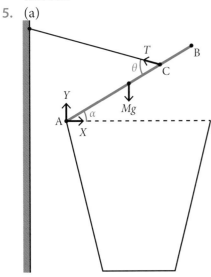

(b) Moments about A:

$$Mg \left(\frac{l}{2} \right) \cos 30 = T \left(\frac{3}{4} l \right) \sin 45$$

$$\frac{Mgl\sqrt{3}}{4} = 40 \times \frac{3}{4} l \times \frac{\sqrt{2}}{2}$$

$$\frac{Mg\sqrt{3}}{4} = 15\sqrt{2}$$

$$Mg = \frac{60\sqrt{2}}{\sqrt{3}} = 20\sqrt{6}$$

$$M = \frac{20\sqrt{6}}{g} = 5.0 \text{ kg (2 s.f.)}$$

(c) $T = \dfrac{2Mg \cos \alpha}{3 \sin \theta}$

(d) $T = \dfrac{2Mg \cos \alpha}{3 \sin \theta}$

$$= \frac{2(10\sqrt{6}) \cos \alpha}{3 \times \frac{\sqrt{2}}{2}} \approx 23.1 \cos \alpha$$

The maximum value of $\cos \alpha$ is 1, so the maximum tension is 23.1 N. The string will not break.
(e) The bin lid is modelled as a uniform rod; the lid is smoothly hinged to the bin; the string is light and inextensible.

6. (a)

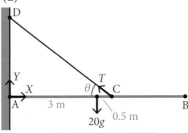

(b) $R = \sqrt{(235.2)^2 + (28)^2}$ = 237 N (3 s.f.) (c) 6.8° (1 d.p.)

7. (a) Resolving in the horizontal:

$$2g = \frac{\sqrt{2}}{2} T \cos 45$$

$$2g = \frac{\sqrt{2}}{2} T \left(\frac{\sqrt{2}}{2} \right) = \frac{T}{2}$$

$$T = 4g \text{ N}$$

Moments about A gives $M = 3$ kg
(b) $Y_1 = g$ N
(c) $X = 2\sqrt{3}g$; $Y = g$

Exercise 6A

1. 24.3 m s^{-1}
2. $(128\mathbf{i} + 23\mathbf{j})$ m

Exercise 6B

1. Car: Momentum
   = 1500 × 30 = 45000 kg m s^{-1}
   Ship: Momentum
   = 500 000 × 0.01 = 5000 kg m s^{-1}
   The car has greater momentum.
2. −0.002 kg m s^{-1}
3. $(720\mathbf{i} + 900\mathbf{j})$ m s^{-1}
4. $(-\mathbf{i} + 4\mathbf{j})$ m s^{-1}
5. 0.002 kg or 2 grams
6. 2.36 m s^{-1}

Exercise 6C

1. Y moves with speed 2.5 m s^{-1} to the right.
2. Particle A moves with speed 6 m s^{-1} to the left; B moves with speed 2 m s^{-1} to the right.
3. (a) 3.2 m s^{-1} (b) 3.73 m s^{-1}
4. ⁵⁄₃ kg
5. $x = 1.5, y = 3.5$

Exercise 6D

1. 0.5 m s^{-1}
2. (a) (i) 3.13 m s^{-1} (3 s.f.)
   (ii) 1.04 m s^{-1} (3 s.f.)
   (b) The balls are modelled as particles; the string is light and inextensible.
3. (a) $(50 \times 6) = (50 \times v) + 30(v + 2)$
   $\Rightarrow v = 3$;
   Block A: 3 m s^{-1}; Block B: 5 m s^{-1}
   (b) 3.75 m s^{-1}
4. (a) ¹⁰⁄₃ m s^{-1} to the left
   (b) ⁵⁄₆ m s^{-1} or 0.833 m s^{-1} (3 s.f.)
   (c) The string is light and inextensible; it is assumed that the string does not affect the collision between the spheres.
5. $\left(-\dfrac{10}{7}\mathbf{i} + \dfrac{15}{7}\mathbf{j} \right)$ m s^{-1}

Exercise 6E

1. 45 m s^{-1} to the left
2. 7.5 m s^{-1}
3. (a) 30 m s^{-1} (b) The two parts of the TV are being modelled as particles. It is assumed they do not lose any speed as they travel through the air. It is assumed that there are no other pieces.
4. $14\mathbf{i} - 21\mathbf{j}$ m s^{-1}
5. (a) $v^2 = u^2 + 2as$
   $v^2 = 20^2 + 2(-10)(10.2)$
   $v^2 = 196$, giving $v = 14$ m s^{-1}
   (b) The second part of the firework begins moving downwards with a speed of 2.5 m s^{-1}.

Exercise 6F

1. 1.5 m s^{-1}
2. 2 m s^{-1}

3. (a) 4 m s⁻¹ (b) The trucks are modelled as particles; the trucks are moving smoothly on the track.

4. (a) $\frac{u}{2}$ m s⁻¹ (b) $-\frac{u}{6}$ m s⁻¹
   (c) A is stationary and B is now moving to the left, since its velocity is negative.
   (d) $-\frac{1}{9}u$ m s⁻¹

5. (a) $0.8\mathbf{i} - 0.6\mathbf{j}$ km s⁻¹
   (b) $1.2\mathbf{i} - 1.1\mathbf{j}$ km s⁻¹

Exercise 7A

1. (a) 1.125 kg m s⁻¹
   (b) 38700 kg m s⁻¹
2. (a) $v = p + 2$ (b) $p = 2$

Exercise 7B

1. 1.8 N s
2. (a) 4 m s⁻¹ (b) 20 N s (c) 20 N s
   (d) 0.2 s (e) The balls are modelled as particles, the surface is smooth.
3. (a) 6 m s⁻¹ (b) 12 N s
4. (a) 0.8 (b) 0.96 N s
5. (a) 135 m s⁻¹
   (b) 364 500 m s⁻²
   (c) 3645 N (d) 1.35 N s
6. (a) $m_1u_1 + m_2u_2 = m_1v_1 + m_2v_2$
   $(0.08 \times 1.5) + (0.12 \times 1)$
   $= 0.08U + 0.12V$
   $0.24 = 0.08U + 0.12V$
   $\Rightarrow 42 = 14U + 21V$ (1)
   $\frac{U}{V} = \frac{21}{26} \Rightarrow 26U = 21V$ (2)
   Sub for $21V$ in (1):
   $42 = 40U \Rightarrow U = 1.05$
   From (1):
   $V = \frac{42 - 14U}{21} = 1.3$
   (b) 0.036 N s
7. (a) –1 m s⁻¹ (b) 3 or 5 kg
8. (a) –10 kg m s⁻¹ (b) 2000 N
9. (a)
   $s = 7.2$; $u = 0$; $v = ?$ $a = 10$
   $v^2 = u^2 + 2as$
   $v^2 = 2(10)(7.2)$
   $v^2 = 144$
   $v = 12$ m s⁻¹
   (b) 10 m s⁻¹ (c) 29700 N
10. (a) 9 m s⁻¹ (b) 810 N

Exercise 7C

1. (a) $(-90\mathbf{i} + 135\mathbf{j})$ N s
   (b) $(31\mathbf{i} - 39\mathbf{j})$ m s⁻¹
2. (a) $(5\mathbf{i} + 19.5\mathbf{j})$ m s⁻¹
   (b) 37.5$\mathbf{i}$ N s
3. (a) $m_X\mathbf{u}_X + m_Y\mathbf{u}_Y = m_X\mathbf{v}_X + m_Y\mathbf{v}_Y$
   $2(2\mathbf{i} + 3\mathbf{j}) + 1(\mathbf{i} + 1.5\mathbf{j})$
   $= 2(p\mathbf{i} + 2\mathbf{j}) + 1\left(\frac{7}{3}\mathbf{i} + q\mathbf{j}\right)$
   $\Rightarrow 5\mathbf{i} + 7.5\mathbf{j} = \left(2p + \frac{7}{3}\right)\mathbf{i} + (4 + q)\mathbf{j}$
   Equating $\mathbf{i}$ components:
   $2p + \frac{7}{3} = 5 \Rightarrow p = \frac{4}{3}$
   Equating $\mathbf{j}$ components:
   $4 + q = 7.5 \Rightarrow q = 3.5$
   (b) $\left(\frac{4}{3}\mathbf{i} + 2\mathbf{j}\right)$ N s
4. (a) 26 m s⁻¹
   (b) $\mathbf{v} = 30\mathbf{i} - 54\mathbf{j}$ m s⁻¹
5. (a) $\mathbf{v} = (6t - 6)\mathbf{i} + (9t^2 - 4)\mathbf{j}$ m s⁻¹
   (b) $t = \frac{2}{3}$ (c) $\mathbf{v} = (4\mathbf{i} - 7\mathbf{j})$ m s⁻¹
6. (a) $\mathbf{v} = (6t^2 - 10t + 1)\mathbf{i} + \left(\frac{2}{3}t^3 - 2t^2 - 4\right)\mathbf{j}$ m s⁻¹
   (b) 25 m s⁻¹

Exercise 8A

1. (a) $\frac{16}{32} \times \frac{15}{31} = \frac{15}{62}$
   (b) $2\left(\frac{16}{32} \times \frac{16}{31}\right) = \frac{16}{31}$
   (c) $\frac{8}{32} \times \frac{7}{31} \times \frac{6}{30} = \frac{7}{620}$
2. (a) Description 2
   (b) Description 3
   (c) Description 1
3. (a) (i) 0.2 (ii) 0.3 (iii) 0.65 (b) 0.15
4. (a) 0.45 (b) 0.85 (c) 0.3 (d) 0.9

Exercise 8B

1. (a)

	School dinner	Packed lunch	Total
Male	15	7	22
Female	18	10	28
Total	33	17	50

(b) (i) ¹¹⁄₂₅ (ii) ⁷⁄₂₂ (iii) ⁷⁄₁₇ (iv) ⁹⁄₁₄

2. (a)

	Annual membership	Pay per visit	Total
Male	40	15	55
Female	43	22	65
Total	83	37	120

(b) (i) ¹¹⁄₂₄ (ii) ⁴⁰⁄₅₅ = ⁸⁄₁₁ (iii) ²²⁄₃₇

3. (a)

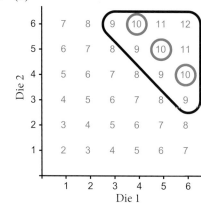

(b) ³⁄₁₀

4. (a)

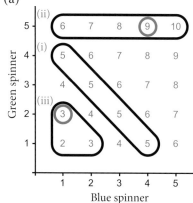

(b) (i) $P(X = 5) = \frac{4}{25}$
(ii)
$P(X = 9 \mid \text{Green spinner is 5}) = \frac{1}{5}$
(iii)
$P(\text{Green spinner is 2} \mid X < 4) = \frac{1}{3}$
(c) The spinners are both fair, i.e. each number is equally likely to come up.

Exercise 8C

1. (a)

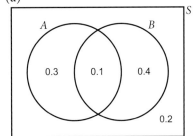

(b) 0.2 (c) 0.7 (d) ⅔

2. (a)

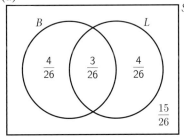

(b) ³⁄₇

3. (a)

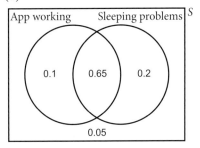

(b) ¹³⁄₁₅

4. (a)

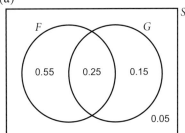

(b) (i) 0.95 (ii) 0.625 (iii) ⁵⁄₁₆
(iv) 0.25

5. (a) 0.16 (b) (i) ⁵⁄₁₈ (ii) ³⁸⁄₆₇
(iii) ⁷⁄₅₃ (iv) ²²⁄₄₅

6. (a) (i) ¹¹⁄₃₁ (ii) ⁵⁄₃₁

(b) $P(H) = \dfrac{11}{31}; P(N) = \dfrac{10}{31};$

$P(H) \times P(N) = 0.114$ (3 s.f.)

$P(H \cap N) = \dfrac{3}{31} = 0.0968$ (3 s.f.)

Playing hockey and playing
netball are not independent

events.

7. $a = 0.32; b = 0.08; c = 0.12$

Exercise 8D

1. (a) ¹³⁄₁₁₀ (b) ⁴³⁄₅₅

(c) $P(B'|A) = \dfrac{P(B' \cap A)}{P(A)}$

$= \dfrac{11/110}{24/110} = \dfrac{11}{24}$

(d) $P(A) \times P(B) = \dfrac{24}{110} \times \dfrac{1}{2}$

$= \dfrac{6}{55} \neq P(A \cap B)$

Comment: one of the models may
be more reliable than the other.

2. (a) Let M and W be the events
that the man and the woman
carry donor cards respectively.
$P(M \cap W) = P(M|W) \times P(W)$

$= \dfrac{4}{9} \times \dfrac{2}{3} = \dfrac{8}{27}$

(b) ³¹⁄₅₄ (c) ⁷⁄₅₄

3. (a) 0.036 (b) 0.344 (c) 0.18
4. (a) 0.6 (b) 0.9 (c) 0.5
5. (a)

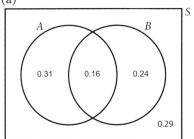

(b) $P(A) = 0.47; P(B) = 0.4$
(c) $P(A|B') = 0.31$
(d) $P(A) \times P(B) = 0.47 \times 0.4$
$= 0.188; P(A \cap B) = 0.16$
$P(A) \times P(B) \neq P(A \cap B)$
Therefore not independent.

6. ¹⁄₈₀
7. (a)

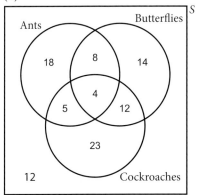

(b) ²³⁄₉₆ (c) ¹¹⁄₁₆ (d) ⁴⁹⁄₆₆
(e) Let B be the event a paper
relates to butterflies. Let C be
the event a paper relates to
cockroaches.

$P(B) = \dfrac{8 + 14 + 4 + 12}{96} = \dfrac{19}{48};$

$P(C) = \dfrac{5 + 4 + 12 + 23}{96} = \dfrac{11}{24}$

$P(B) \times P(C) = \dfrac{19}{48} \times \dfrac{11}{24} = \dfrac{209}{1152}$

$P(B \cap C) = \dfrac{4 + 12}{96} = \dfrac{1}{6}$

$P(B) \times P(C) \neq P(B \cap C)$
Therefore events B and C are not
independent.

Exercise 8E

1. (a)

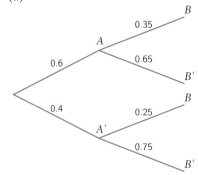

(b) (i) 0.21 (ii) 0.1 (iii) 0.31
(iv) ²¹⁄₃₁ = 0.677 (3 s.f.)

2. (a)

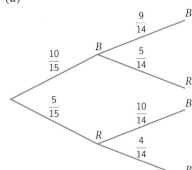

(b) (i) ⅓ (ii) ²⁄₇

3. (a)

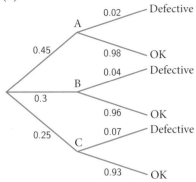

(b) $\dfrac{9}{1000}$ or 0.009

(c) $\dfrac{77}{2000}$ or 0.0385

(d) $5/11$ or 0.455 (3 s.f.)

4. $51/62$

5. (a)

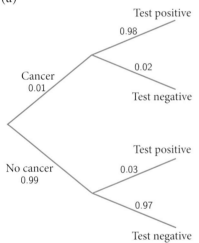

$P(\text{positive}) = (0.01 \times 0.98)$
$+ (0.99 \times 0.03) = 0.0395$

(b) $P(\text{cancer}|\text{positive})$

$= \dfrac{P(\text{cancer} \cap \text{positive})}{P(\text{positive})}$

$= \dfrac{0.01 \times 0.98}{0.0395}$

$= 0.248$ (3 s.f.) $\approx 25\%$

(c) Considering the accuracy of the test (98% correct for patients with the disease and 97% correct for patients without), it may be a surprise that, given a positive test, there is only a 0.248 chance that the patient has the disease. This occurs because of the false positives. Imagine testing 100 people, 1 with the disease and 99 without. Assume the test correctly picks up the disease in

the one person with it. Out of the 99 without, there will be about 3 false positives. In total there are four positive tests, only one of which is accurate.

The lessons to learn from this are: (1) It may not be appropriate to test everybody over 60. Only testing those who are at greatest risk or showing symptoms may be more appropriate. (2) If a patient tests positive as a part of a mass testing regime, they should have a second test to confirm the result.

6. (a) p^2 (b) $(1 - p)^2$
    (c) $p^2 + (1 - p)^2$
    (d) $P(\text{HH}|\text{Both same})$

$= \dfrac{P(\text{HH} \cap \text{Both same})}{P(\text{Both same})}$

$= \dfrac{P(\text{HH})}{P(\text{Both same})}$

$= \dfrac{p^2}{p^2 + (1 - p)^2}$

(e) $\dfrac{p^2}{p^2 + (1 - p)^2} = 0.2$

$p^2 = 0.2(2p^2 - 2p + 1)$

$p^2 - 0.4p^2 + 0.4p - 0.2 = 0$

$0.6p^2 + 0.4p - 0.2 = 0$

$3p^2 + 2p - 1 = 0$

$(3p - 1)(p + 1) = 0$

$3p - 1 = 0 \Rightarrow p = \dfrac{1}{3}$

or $p + 1 = 0 \Rightarrow p = -1$ (reject)

Exercise 9A

1. $a = 8$; $b = 2$
2. (a) 0.01 (b) 0.18
3. (a) 25 mph (b) 15 mph
    (c) (i) The mean speed is lower on Monday as the traffic is moving with a lower average speed.
    (ii) The standard deviation is lower on Monday as the spread of the speeds is lower.
    (d) There is more traffic on the road at 8:30 on a Monday morning as people go to work and school, etc. The lower mean occurs because the road

is congested, and the traffic is generally slower. The lower standard deviation occurs because the cars must drive at similar speeds, so the spread of speeds is lower.

Exercise 9B

1.

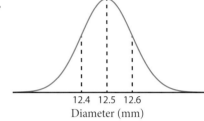

2. $\mu = 108$ g

3. $\mu = 7$; $\sigma = 2$

Exercise 9C

1. (a) 0.894 (3 s.f.) (b) 0.994 (3 s.f.)
    (c) 1
2. (a) 0.0668 (3 s.f.)
    (b) 0.00135 (3.s.f)
    (c) 1 (d) 0.683 (3.s.f)
3. (a) (i) 0.2660 (ii) 0.7340
    (b) $P(Q < 80) + P(Q > 80) = 1$. The total area under the curve is 1.
4. (a) (i) 0.212 (ii) 0.0548
    (b) The question states that the recorded temperatures **roughly** followed a normal distribution. Therefore, any probabilities calculated using a normal distribution model are estimates.
5. (a) 0.0572 (b) 0.295
6. (a) (i) 0.268 (ii) 0.0599 (b) 18

Exercise 9D

1. (a) 62.1 (b) 62.1 (c) 59.0
2. (a) 9.81 (b) 10.1 (c) 10.8
3. (a) 83.5 g (b) 62.7 g (c) 70.3 g
4. 10:20 am
5. (a) 60% (b) 72%
6. 38
7. (a) Q_2 is the median value. In a normal distribution, the median is equal to the mean, since the distribution is symmetrical. The mean is 190 cm, therefore $Q_2 = 190$ cm. (b) 20.2 cm

8. (a) (i) 0.908 (ii) 0.0918
(iii) 0.160 (iv) 0.0228
(b) 110.1 grams

Exercise 9E

1. (a) 0.9821 (b) 0.9147
(c) 0.0548 (d) 0.2266
2. (a) 0.0082 (b) 0.0571
(c) 0.4960 (d) 0.2148
3. (a) 0.1359 (b) 0.0700
(c) 0.1972 (d) 0.0250
4. (a) 0 (b) 1 (c) –1.75 (d) –0.5
5. (a) $\Phi(1.25)$ (b) $\Phi(-0.5)$
(c) $1 - \Phi(0.5)$
(d) $\Phi(0.75) - \Phi(-0.25)$
6. (a) (i) $a = 1.18$ (ii) 0.119 (b) 90%
7. (a) Roughly $z = -0.674$ to
$z = 0.674$
(b) (i) $R < 802$ mm (3 s.f.)
(ii) $R > 978$ mm (3 s.f.)

Exercise 9F

1. 9.8
2. 4
3. 120.1 N
4. 12.5
5. 37.76
6. 238
7. $\mu = 499.1$ grams; $\sigma = 7.062$ grams
8. $\sigma = 2.5$ minutes;
$\mu = 24.075$ minutes (or
24 minutes 4.5 seconds)
9. (a)

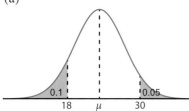

(b) $\mu = 23.26$ km; $\sigma = 4.101$ km
(c) 0.336 (3 s.f.)
10. (a) $\mu = 16.79$ kW h;
$\sigma = 0.9420$ kW h
(b) 1.27 kW h

Exercise 10A

1. ⅜
2. Bell-shaped curve with horizontal
asymptotes left and right. The
mean value is at the centre,
corresponding to the greatest
probability.

3. There is a fixed number of trials.
The trials are independent. There
are two mutually exclusive and
exhaustive outcomes for each
trial. There is a fixed probability
of success for each trial.
4. (a) $X \sim N(80, 6^2)$ (b)

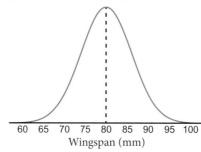

Wingspan (mm)

5. $t = 10 \therefore h^2 = 2.5(10) + 2 = 27$
Model predicts $h = \sqrt{27} = 5.196$
Yes, the model is suitable. The
model predicts the tree will be
5.2 m high (1 d.p.), which is close
to the true height.

Exercise 10B

1. (a) The normal distribution is
widely seen in nature and in
natural processes. (b) Heights
or masses of people or animals
in a population, etc; masses of
a particular grocery item or
factory-produced item, etc.
2. (a) Continuous (b) Discrete
(c) Continuous (d) Discrete
3. (a) Binomial (b) Normal
(c) Binomial (d) Normal
(e) Normal
(f) Neither. X is non-numeric.
4. (a) X can be modelled as a
binomial random variable.
• There is a fixed number of
trials, 4; • There is a fixed
probability of 'success' (choosing
white), ³⁄₁₀; • There are two
possible mutually exclusive and
exhaustive outcomes (white and
black) for each trial. • Trials are
independent.
(b) Neither. X cannot be
modelled as a binomial random
variable because there is not a
fixed probability of choosing
white, as the number of beads
decreases. X cannot be modelled

as a normal random variable
since it is discrete.
5. Neither. X is not normal, since
this is a discrete distribution (X
can take only the values 1, 2, 3,
4, 5 and 6.) However, it cannot
be modelled using the binomial
distribution either. One way to
spot this is: $P(X = 1) = P(X = 2)$
$= P(X = 3) = P(X = 4) = P(X = 5)$
$= P(X = 6) = ⅙$. This is known
as a uniform distribution. In
a binomial distribution the
probabilities are not all equal.
[Note: a detailed study of the
uniform distribution was not a
part of the CCEA specification at
the time of publication.]
6. Each trial has a pair of mutually
exclusive, exhaustive outcomes:
'effective' and 'not effective'. The
probability of effectiveness of
the drug is the same between
patients. The outcomes for each
patient are independent of each
other.
7. (a)(i) Binomial, since a fixed
number of trials, 20; trials
independent; two possible
outcomes for each trial ('lays' and
'doesn't lay'); fixed probability of
each hen laying. (ii) Neither. Y
cannot be normal as it is discrete.
It is not binomial as, in theory,
there is no upper limit to the total
number of eggs.
(b) Assumptions for (a)(i): there
is a fixed probability of each hen
laying. This may not be the case
if, for example, some are older,
different breeds, etc.
8. No, the normal distribution is
not a suitable model for the data
because the distribution is not
symmetrical.
9. (a) The normal distribution
is a symmetrical distribution,
which means that the median
is the same as the mean.
$\mu = 4 \therefore Q_2 = 4$ m
(b)(i) The mean increases
when some of the smaller trees

are removed. (ii) With some of the smaller trees removed, the distribution may become asymmetrical for a time, so a normal model may become inappropriate. The population may revert to a normal distribution over time.

10. Neither. The probability distribution for X has a similar shape to the normal distribution, but there are key differences, for example: • X is a discrete random variable, since only certain values of X are possible. The normal distribution is continuous. • For a normally distributed random variable, extreme values are possible. In this case, X cannot take a value less than 1 or greater than 6.
X is not a binomial random variable, because, for example • its value does not arise from counting 'successes' in a binomial trial; • the possible values of X in a binomial distribution are always integer values.

Exercise 10C

1. The coin is fair.
2. (a) Siobhán has made the assumption that her dice are all fair, i.e. that $P(6) = \dfrac{1}{6}$ for each one of her 12 dice.
   (b) There is always some uncertainty in the prediction of a model. So, it is not possible for Siobhán to be certain about the number of heads and tails, even if her modelling assumption that the dice are all fair is correct.
3. (a) $\dfrac{1}{16}$ (b) It is assumed that the probability Carol has a girl is exactly 0.5 every time she has a baby.
4. Deirdre's modelling assumption was that all books on the shelf are equally likely to be chosen. This may not be valid: larger books may be more likely to be chosen

when Deirdre picks one with her eyes closed.

5. (a) 0.55 or $\dfrac{11}{20}$
   (b) The model does not consider the fact that there is usually more traffic on the roads each weekday, and less at the weekend.
6. (a) Aoife is assuming the spinner is fair, in that the probabilities are proportional to the sizes of the angles. The blue section has an angle of 40°. Since this is $\dfrac{1}{9}$ of 360°, so $P(\text{blue}) = \dfrac{1}{9}$.
   (b) (i) $P(\text{red}) = \dfrac{2}{9}$
   (ii) $P(\text{red}|\text{not blue}) = \dfrac{1}{4}$
7. (a)

Length of queue in front of Dan (number of people)	Number of days	Relative frequency (estimate of probability)
0	24	$\dfrac{24}{50} = \dfrac{12}{25}$
1	18	$\dfrac{18}{50} = \dfrac{9}{25}$
2	6	$\dfrac{6}{50} = \dfrac{3}{25}$
>2	2	$\dfrac{2}{50} = \dfrac{1}{25}$

   (b) $200 \times \dfrac{1}{25} = 8$
   (c) $0.1 \times \dfrac{21}{25} + 0.5 \times \dfrac{4}{25}$
   $= 0.164 = 16.4\%$
   (d) (i) The probabilities used are estimates, so his estimate for the probability of being late for work is also an estimate. (ii) The modelling could be improved by using more data, i.e. surveying a larger number of days. (iii) He could • leave his house 5 or 10 minutes earlier; • avoid joining the queue if there are, for example, two or more people

in front of him; • try a different coffee shop; or • stop getting coffee on the way to work!

8. (a) 0.42 (b) $P(\text{not scoring})$
   $= P(\text{saved}) + P(\text{missing})$
   $= 0.42 + 0.15 = 0.57$
   (c) 0.0410 or 4.10% (3 s.f.)
   (d) Possible assumptions: • A fixed probability 0.7 of saving a penalty, given diving the correct way. • Players have not changed their technique since research undertaken. • Players line up in expected way – no unexpected substitutes, etc.
9. (a)

   (b) (i) $P(X = 7) = \dfrac{6}{36} = \dfrac{1}{6}$
   (ii) $P(X = 10 \mid \text{Black die is 6}) = \dfrac{1}{6}$
   (iii) $P(X = 3 \mid X \leq 5) = \dfrac{2}{10} = \dfrac{1}{5}$
   (c) Both dice are fair, i.e. each number is equally likely to come up, with a probability of $\dfrac{1}{6}$.
10. (a) $\dfrac{1}{8}$ (b) The die is biased towards 6: $P(X = 6) = 0.5$
11. (a) Based on snow falling 3 times in 30 years, $p = \dfrac{3}{30} = 0.1$
   The model gives $p^2 = \dfrac{5 - 4.1}{100}$
   $= 0.009 \Rightarrow p = 0.0949$ (3 s.f.), which is very close to the observed probability. We conclude that, based on this data, the model is suitable.
   (b) For temperatures above 5°C the probability is undefined. For

temperatures below –95°C the predicted probability becomes greater than 1.
(c) Possible answers include: air pressure, cloud cover, temperature at higher level.

Exercise 11A

1. (a) ⅛ (b) ⅜
2. (a) Testing whether a sofa catches fire may be a destructive test – any sofa tested will be unsellable, so as small a sample as possible should be used. (b) It would be too costly and time consuming to ask every person. (c) Oak trees have thousands of leaves – it would be impossible to measure all of them.

Exercise 11B

1. (a) A statistical hypothesis is a statement made about a population parameter that we test using evidence from a sample. (b) The null hypothesis is what we assume to be correct and the alternative hypothesis is what we conclude if our assumption is wrong. (c) H_0 and H_1
2. (a) A one-tailed test is used to determine whether the parameter being tested has differed in one direction only (either upwards or downwards). (b) Right-sided (for example testing $\mu > 8$) and left-sided (testing $\mu < 8$).
3. 0.1
4. (a) Evidence is gained by taking a **sample**. (b) The data from the sample is summarised in some way to calculate the **test statistic**. It may, for example, be the mean weight in a sample of 50 boxes.
5. The test statistic is the number of accidents at the intersection. Note: no sample has been taken here. The change in the test statistic will be investigated by analysis of all the data available. Alternatively, a sample of the data could be taken, for example the

number of accidents taking place on the first day of each month.
6. (a) The test statistic is N, the number of sixes. (b) One-tailed, right-sided (since Farooq is testing whether the proportion of sixes is greater than a particular value) (c) $H_0: p = \dfrac{1}{6}$, where p is the probability of getting a 6.
(d) $H_1: p > \dfrac{1}{6}$
7. (a) The test statistic is N, the number of times she gets a head.
(b) Two-tailed (c) $H_0: p = \dfrac{1}{2}$, where p is the probability of getting a head.
(d) $H_1: p \neq \dfrac{1}{2}$
8. (a) The test statistic is the number of accidents in a given month or other specified time period.
(b) Two-tailed (c) $H_0: \lambda = 5$
(d) $H_1: \lambda \neq 5$
9. (a) A suitable test statistic is p the proportion of faulty articles in a batch. (b) One-tailed, left-sided, since the test is to determine whether there is a **reduction** in the proportion, i.e. whether p is less than a certain value.
(c) $H_0: p = 0.15$; $H_1: p < 0.15$
(d) The null hypothesis is rejected if the test shows, at the 5% significance level, that the proportion is less than 0.15.
10. (a) z_{test}
(b) $H_0: \mu = 7.8$; $H_1: \mu < 7.8$
(c) A one-tailed, left-sided test.

Exercise 12A

1. 0.238 (3 s.f.)
2. Evidence is collected in the form of a sample. The test statistic is the quantity that is under investigation and is derived from the sample. The test statistic is used to determine whether to reject the null hypothesis.
3. A two-tailed test is required, because we are testing for an increase or decrease in the mean.

Exercise 12B

1. (a) If a sample is taken from a population, the sample mean is the mean of that sample. (b) For a single sample, the sample mean is not necessarily equal to the population mean.
2. (a) $\bar{X} \sim N(400, 40)$ (b) An increase in the sample size results in a decrease in the standard deviation (and variance) of the distribution of the sample mean.
3. (a) $X \sim N(35, 42.25)$
(b) $\bar{X} \sim N(35, 4.225)$
(c) The assumption is that the distribution of the hotel population does not change over time.
4. (a) $M \sim N(550, 45^2)$
(b) (i) $\bar{M} \sim N(550, 202.5)$
(ii) $\bar{M} \sim N(550, 20.25)$

Exercise 12C

1. (a) $z_{test} = 0.958$; $z_{crit} = \pm 1.960$. Cannot reject H_0.
(b) $z_{test} = -5.52$; $z_{crit} = -1.282$. Reject H_0; accept H_1.
(c) $z_{test} = 5.48$; $z_{crit} = 1.960$. Reject H_0; accept H_1.
(d) $z_{test} = 1.28$; $z_{crit} = \pm 2.326$. Cannot reject H_0.
2. (a) $H_0: \mu = 14.7$; $H_1: \mu > 14.7$
$$z_{test} = \frac{15.1 - 14.7}{\sqrt{2.2^2/50}} = 1.286 \text{ (3 d.p.)}$$
For a one-tailed test at a 10% level of significance, $z_{crit} = 1.282$ $|z_{test}| > |z_{crit}|$ so we can reject H_0 and accept H_1. We conclude that, at a 10% significance level, the scores are higher at Waterloo Road. (b) The scores of students follow a normal distribution
3. (a) $H_0: \mu = 24$; $H_1: \mu < 24$
(b) One-tailed, left-sided
(c) $z_{test} = \dfrac{20 - 24}{\sqrt{2^2/50}}$
$= -14.1$ (3 s.f.)
(d) $z_{crit} = -1.645$
(e) Yes. Since $|z_{test}| > |z_{crit}|$, reject H_0 and adopt H_1. At a significance level of 5%, there is

sufficient evidence to conclude that the passengers were in the water for less than 24 minutes.

4. (a) $H_0: \mu = 21.2$; $H_1: \mu > 21.2$
(b) One-tailed (right-sided)
(c) $z_{test} = \dfrac{22.1 - 21.2}{\sqrt{5^2/100}} = 1.8$
(d) $z_{crit} = 1.645$
(e) Yes. Since $|z_{test}| > |z_{crit}|$, they can reject H_0 and accept H_1. They conclude there is sufficient evidence at the 5% significance level that temperatures have increased.
(f) No. They have shown they are 95% confident that temperatures have increased, but cannot ascribe it with certainty to climate change, although meteorologists often use terminology such as 'consistent with climate change'.

5. $X \sim N(950, 100^2)$
$$z_{test} = \frac{\bar{x} - \mu}{\sqrt{\sigma^2/n}}$$
$$z_{test} = \frac{1000 - 950}{\sqrt{100^2/50}} = 3.54$$
$z_{crit} = 1.96$ for a two-tailed test with a 5% significance level
$|z_{test}| > |z_{crit}|$
∴ reject H_0, accept H_1
Patricia concludes that there is sufficient evidence to suggest that the concentrations of bacteria in the river have been affected by the storm.

6. (a) 'Affects' implies non-directional change, so test is two tailed.
(b) $X \sim N(200, 30^2)$
$H_0: \mu = 200$; $H_1: \mu \neq 200$
5% level of significance, two-tailed test, so $z_{crit} = 1.96$
$$z_{test} = \frac{\bar{x} - \mu}{\sqrt{\sigma^2/n}} = \frac{205 - 200}{\sqrt{30^2/120}} = 1.83$$
Since $|z_{test}| < |z_{crit}|$ we do not reject H_0 and conclude that there is insufficient evidence at the 5% level of significance to suggest that drinking coffee before the game affects your score.

(c) 5% or 0.05

7. $X \sim N(8, 4.41)$
$H_0: \mu = 8$; $H_1: \mu < 8$
5% level of significance, one-tailed test, so $z_{crit} = -1.645$
$$z_{test} = \frac{\bar{x} - \mu}{\sqrt{\sigma^2/n}} = \frac{7.75 - 8}{\sqrt{4.41/170}}$$
$$= -1.55$$
Since $|z_{test}| < |z_{crit}|$ we do not reject H_0. We conclude that there is insufficient evidence at the 5% level of significance to suggest that workers are producing a pallet of drinks more quickly due to the bonus scheme.

8. (a) $X \sim N(120, 30^2)$
$H_0: \mu = 120$; $H_1: \mu \neq 120$
Two-tailed test. For a two-tailed test at 10% significance, $z_{crit} = \pm 1.645$
$$z_{test} = \frac{128 - 120}{\sqrt{30^2/50}} = 1.886$$
Since $|z_{test}| > |z_{crit}|$ reject H_0 and accept H_1. We conclude that there is sufficient evidence to show that the mean time is different.
(b) Weather conditions; different profile of competitors, e.g. fewer club runners or more older people, etc.

9. (a) $X \sim N(50, 0.5^2)$
$$\bar{X} \sim N\left(50, \frac{0.5^2}{40}\right)$$
$H_0: \mu = 50$; $H_1: \mu < 50$
One tailed test at the 5% level of significance.
$$z_{test} = \frac{49.97 - 50}{\sqrt{0.5^2/40}} = -0.379$$
For a one-tailed test at a 5% level of significance, $z_{crit} = -1.645$
Since $|z_{test}| < |z_{crit}|$ do not reject H_0. There is insufficient evidence that Peter is getting less than 50 litres of fuel.
(b) The sample standard deviation was used in the calculations. An assumption has been made that this is a good approximation for the population standard deviation.

10. (a) $H_0: \mu = 9.60$; $H_1: \mu > 9.60$
1% level of significance, one-tailed test, so $z_{crit} = 2.326$. If X is the random variable 'the price of items on the auction site', then
$X \sim N(9.60, 1.20^2)$
$$\bar{X} \sim N\left(9.60, \frac{1.20^2}{30}\right)$$
$$z_{test} = \frac{10.2 - 9.6}{\sqrt{1.2^2/30}} = 2.74 \text{ (3 s.f.)}$$
$|z_{test}| > |z_{crit}|$ ∴ reject H_0 and accept H_1. Chris can conclude that there is sufficient evidence, at a 99% confidence level, that the items he has bought on the site this year are more expensive.
(b) Chris may have shopped for a slightly different range of items during the last year (better quality for example). It may not be a result of price inflation on the site.

Exercise 13A

1. (a) 0.239 (3 s.f.) (b) 0.966 (3 s.f.)
2. (a) 0.200 (b) 0.00159 (c) 0.803
3. Two-tailed, since the test is to determine whether the population has 'changed' (not increased or decreased).

Exercise 13B

1. (a) One-tailed, right-sided
(b) One-tailed, left-sided
(c) Two-tailed
2. (a) The critical region is a region of the probability distribution which, if the test statistic falls within it, would cause you to reject the null hypothesis.
(b) The acceptance region is the region of the probability distribution in which we accept the null hypothesis.
(c) The critical value is the first value to fall inside the critical region.
3. 0.0787
4. (a) $H_0: p = 0.15$; $H_1: p > 0.15$
One-tailed
(b) The critical region is 7, 8, 9, …, 20 (c) 7 (d) 0.0219

5. (a) 0.25
   (b) $H_0: p = 0.25$; $H_1: p > 0.25$
   (c) 15, 16, 17, …, 40 (d) 0.0544
6. (a) The test statistic is the number of pea plants that survive.
   $H_0: p = \frac{1}{3}$; $H_1: p > \frac{1}{3}$
   (b) Critical region is $X \geq 17$
   (c) 5.84%
7. (a) The test statistic is X, the number of times Hardeep gets a 5. (b) $H_0: p = 0.2$; $H_1: p > 0.2$
   (c) $X \geq 8$

Exercise 13C

1. (a) $P(X \leq 3) = 0.559 > \alpha$ ∴ Do not reject H_0.
   (b) $P(X \geq 15) = 0.416 > \alpha$ ∴ Do not reject H_0.
   (c) $P(X \leq 2) = 0.0106 < \frac{\alpha}{2}$ ∴ Reject H_0.
2. (a) The test should be two-tailed. She is suggesting that the proportion is not 35%, so she believes it is either higher or lower than this.
   (b) $H_0: p = 0.35$; $H_1: p \neq 0.35$
   (c) Let X be random variable 'The number of cars failing to meet emissions standards in the sample of 15'. Then H_0 is rejected if either $X \leq 1$ or $X \geq 9$.
   (d) 0.0564
3. $H_0: p = 0.2$; $H_1: p > 0.2$
   One-tailed test
   Under H_0, $X \sim B(10, 0.2)$
   $P(X \geq 3) = 1 - P(X \leq 2)$
   $= 1 - 0.6778 = 0.3222$
   The probability is greater than the significance level of 0.1.
   $P(X \geq 3) > 0.1$ ∴ H_0 cannot be rejected. There is insufficient evidence at the 10% significance level that the team's results have improved. The manager is fired.
4. (a) One-tailed (b) H_0: The coin is not biased, so $p = 0.5$
   H_1: The coin is biased in favour of tails, so $p > 0.5$ (c) Under the null hypothesis, $P(X \leq 6) = 0.0577$. Since $P(X \leq 6) > 0.05$, there is insufficient evidence to reject the

null hypothesis at the 5% level. Finn cannot conclude that the coin is biased.
5. (a) Critical region is $X = 0$ and $X \geq 8$ (b) 4.36% using tables or 4.37% from calculator (c) $X = 8$ is in the critical region. There is enough evidence to reject H_0. The hospital's proportion of complications differs from the Northern Ireland figure.
6. (a) 0.972 (3 s.f.) (b) $H_0: p = 0.3$; $H_1: p > 0.3$; $X \sim B(10, 0.3)$
   One tailed test. Reject H_0 if, under H_0, $P(X \geq 6)$ is less than the significance level of 5%.
   $P(X \geq 6) = 1 - P(X \leq 5)$
   $= 1 - 0.9527 = 0.0473$
   Under H_0 there is roughly a 4.73% chance that 6 of the students would reach the required level. There is sufficient evidence, at the 5% level, to reject the null hypothesis.
   (c) The head of department could take a larger sample of students.
7. $H_0: p = 0.1$; $H_1: p > 0.1$
   One-tailed test.
   Under H_0, $X \sim B(50, 0.1)$
   $P(X \geq 11) = 1 - P(X \leq 10)$
   $= 1 - 0.9906 = 0.0094$
   Under H_0, the probability of 11 customers out of 50 choosing the wholegrain loaf is 0.0094, which is less than the significance level of 0.01. $P(X \geq 11) < 0.01$ ∴ reject H_0 and accept H_1. The manager concludes that there is evidence, at a 1% significance level, that the wholegrain loaf has become more popular this week.
8. (a) $H_0: p = 0.2$; $H_1: p > 0.2$
   (b) Using $n = 30$ and $p = 0.2$, from the table
   $P(X \leq 9) = 0.9389$
   $\Rightarrow P(X \geq 10) = 0.0611$
   $P(X \leq 10) = 0.9744$
   $\Rightarrow P(X \geq 11) = 0.0256$
   ∴ $X \geq 11$ is the critical region.
   (c) Since 8 is not in the critical region, Charlie cannot reject H_0. There is insufficient evidence at

the 5% level of significance that the proportion of lime sweets has increased.
   (d) The critical region has a probability of 0.0256 or 2.56%.
9. Test statistic: the number of T-shirts with defects.
   $H_0: p = 0.1$; $H_1: p \neq 0.1$
   (two tailed).
   $P(X \leq 1) = 0.3917 = 39.17\%$
   $39.17\% > 5\%$ so there is not enough evidence to reject H_0. There is not enough evidence to suggest the proportion of T-shirts with defects has changed.
10. (a) Two-tailed, since she suspects the proportion is different, not increased or decreased.
    (b) $H_0: p = 0.35$; $H_1: p \neq 0.35$
    (c) Since the significance level is 10%, we require 5% in each tail.
    $P(X \leq 1) = 0.0424$, whereas $P(X \leq 2) = 0.1513$, which is too big. ∴ $X = 0, 1$ are in the lower tail. $P(X \leq 7) = 0.9745$
    $\Rightarrow P(X \geq 8) = 0.0255$
    $P(X \leq 6) = 0.9154$
    $\Rightarrow P(X \geq 7) = 0.0846$, which is too big. ∴ $X = 8, 9, 10, 11, 12$ are in the upper tail.
    Overall: the critical region is 0, 1, 8, 9, 10, 11, 12
    (d) Since 6 is not in the critical region, H_0 cannot be rejected. Kate does not have sufficient evidence to conclude that the proportion of bad apples is different this year.
    (e) Significance level
    $= 0.0424 + 0.0255 = 0.0679$
    or 6.79% (f) Kate could use a larger sample of apples.
11. If X is the number of people in Freya's survey who say 'support the monarchy', we can consider X to follow a binomial distribution, with the probability of success 0.62, assuming the null hypothesis. Test $H_0: p = 0.62$ against $H_1: p \neq 0.62$.
    This is a two-tailed test, since Freya's belief is that the

Queen's death has changed the proportion, not increased it or decreased it. Use a probability of 2.5% or 0.025 in each tail. Under the null hypothesis,

$X \sim B(40, 0.62)$

$P(X \geq 28) = 0.0779$

(Note: this calculation is done on the calculator, since the probability of 0.62 is not in the binomial cumulative distribution tables.) Since $0.0779 > 0.025$, Freya cannot reject H_0. There is insufficient evidence at the 5% level that the proportion of people who support the monarchy has changed.

12. $X \sim B(40, 0.01)$

$H_0: p = 0.01$; $H_1: p > 0.01$

$P(X \geq 2) = 1 - P(X \leq 1)$

$= 1 - 0.9392 = 0.0608$

$P(X \geq 3) = 1 - P(X \leq 2)$

$= 1 - 0.9925 = 0.0075$

The value closest to 0.05 is 0.0608 Therefore, the values in the critical region are $2 \leq X \leq 40$ Since 2 lies in the critical region, the YouTuber can reject H_0. He has sufficient evidence to conclude, at the 5% level of significance, that more people have heard of him now.

Exercise 14A

1. (a) 0.866 (b) The correlation coefficient of 0.866 indicates a strong positive correlation between the scores of the two judges. If one judge gives a high score, the other judge is also likely to give a high score.

2. (a) $H_0: p = 0.75$; $H_1: p > 0.75$
   (b) This is a one-tailed test, since Bernard is testing in one direction only (that the proportion is **greater than** 0.75).
   (c) 10% or 0.1 (d) Bernard should not say he has proved his turkeys are larger. A hypothesis test is not a proof. He has collected evidence to suggest that his turkeys are larger.

Exercise 14B

1. (a) $H_0: \rho = 0$; $H_1: \rho > 0$
   $r_{crit} = 0.6084$; $|r| < r_{crit}$ ∴ cannot reject H_0. There is no evidence $\rho > 0$ at the 10% level of significance.
   (b) $H_0: \rho = 0$; $H_1: \rho \neq 0$
   $r_{crit} = 0.5614$; $|r| > r_{crit}$ ∴ reject H_0 and accept H_1. There is evidence $\rho \neq 0$ at the 1% level of significance.
   (c) $H_0: \rho = 0$; $H_1: \rho < 0$
   $r_{crit} = 0.4409$; $|r| > r_{crit}$ ∴ reject H_0 and accept H_1. There is evidence $\rho < 0$ at the 5% level of significance.
   (d) $H_0: \rho = 0$; $H_1: \rho \neq 0$
   $r_{crit} = 0.1654$; $|r| > r_{crit}$ ∴ reject H_0 and accept H_1. There is evidence $\rho < 0$ at the 10% level of significance.
   (e) $H_0: \rho = 0$; $H_1: \rho > 0$
   $r_{crit} = 0.3120$; $|r| < r_{crit}$ ∴ cannot reject H_0. There is no evidence $\rho > 0$ at the 2.5% level of significance.

2. (a) 0.243 (3 s.f.)
   (b) $H_0: \rho = 0$; $H_1: \rho > 0$
   One-tailed. $n = 9$; $\alpha = 0.1$
   From the table $r_{crit} = 0.4716$
   $|r| < r_{crit}$ ∴ cannot reject the null hypothesis: no evidence that $\rho > 0$. There is no evidence, at the 10% level of significance, that there is a correlation between the height of a castle and the time taken for its flag to fall.

3. (a) $H_0: \rho = 0$ and $H_1: \rho \neq 0$
   Critical r value 0.2787;
   $0.301 > 0.2787$ ∴ reject H_0; accept H_1; evidence of correlation.
   (b) $H_0: \rho = 0$ and $H_1: \rho \neq 0$
   Critical value 0.3281;
   $0.301 < 0.3281$ ∴ do not reject H_0; insufficient evidence of correlation

4. (a) –0.978 (3 s.f.)
   (b) $H_0: \rho = 0$; $H_1: \rho \neq 0$
   Critical r value 0.8343. Since $|r| > r_{crit}$, there is evidence to reject H_0 and accept H_1. There is evidence of correlation between

the number of checks and the number of errors.

5. $H_0: \rho = 0$; $H_1: \rho > 0$
   One-tailed. $n = 10$; $\alpha = 0.05$
   From the table $r_{crit} = 0.5494$
   $|r| < r_{crit}$ ∴ cannot reject the null hypothesis. There is no evidence in this dataset, at the 5% level of significance, that there is a correlation between a baby's age and weight.

6. $H_0: \rho = 0$; $H_1: \rho < 0$
   One-tailed. $n = 7$; $\alpha = 0.01$
   From the table $r_{crit} = 0.8329$
   $|r| > r_{crit}$ ∴ reject the null hypothesis and accept the alternative hypothesis. There is evidence in this dataset, at the 1% level of significance, that there is a correlation between the air temperature and air pressure in Belfast.

7. (a) $r = 0.686$ (3 s.f.).
   (b) $H_0: \rho = 0$; $H_1: \rho \neq 0$
   Two-tailed. $n = 9$; $\dfrac{\alpha}{2} = 0.025$
   From the table $r_{crit} = 0.6664$
   $|r| > r_{crit}$ ∴ reject the null hypothesis and accept the alternative hypothesis that $\rho > 0$. There is evidence, at the 5% level of significance, that there is a correlation between age and salary.

8. (a) If the team is playing well, they will tend to score more goals and let fewer in. If the team is not playing well, they will tend to score fewer goals and let more in. Hence you would expect r to be negative.
   (b) $H_0: \rho = 0$; $H_1: \rho < 0$
   $n = 20$, $r = -0.825$
   From the table, the critical r value $r_{crit} = 0.5155$. $|r| > r_{crit}$ ∴ reject H_0 and accept H_1. There is evidence at the 1% significance level of a negative correlation between the goals for and goals against. Hence r is significant at the 1% level of significance.

9. (a) $r = 0.593$
   (b) $H_0: \rho = 0$; $H_1: \rho > 0$

(c) For $n = 6$ and $\alpha = 0.005$, $r_{crit} = 0.9172$. $|r| < r_{crit}$ ∴ cannot reject H_0. No evidence at 0.5% significance level for a correlation.
(d) She could use a larger sample of patients; she could increase the significance level of her test, for example to 5%.

10. (a) One-tailed (b) For $n = 15$ and $\alpha = 0.05$, $r_{crit} = 0.4409$. $|r| > r_{crit}$ ∴ reject H_0 and accept H_1. There is evidence at the 5% significance level of a correlation between mass and lifespan.